Microwave Amplifiers and Oscillators

Microwave Amplifiers and Oscillators

by

Christian Gentili

Direction des Recherches Études et Techniques, Paris

Translated by

E. Griffin

McGraw-Hill Book Company

New York St. Louis San Francisco Auckland
Bogota Hamburg Johannesburg London Madrid
Mexico Montreal New Delhi Panama Paris
Sao Paulo Singapore Sydney Tokyo Toronto

Revised 1987

English edition first published 1987
by North Oxford Academic Publishers Ltd,
a subsidiary of Kogan Page Ltd, 120 Pentonville Road,
London N1 9JN

Published in the U.S.A. by McGraw-Hill Inc

ISBN 0-07-022995-3

Printed and bound in Great Britain by
Biddles Ltd, Guildford and King's Lynn

Contents

Preface

Over the past decade the gallium arsenide metal–semiconductor field effect transistor not only has taken over from bipolar transistors at frequencies of a few gigahertz but also has presented a serious challenge to travelling wave tubes just as, for example, some oscillator devices using Gunn or IMPATT diodes as active elements. Field effect transistors are now commercially available with noise factors at 18 GHz of only 1.8 dB for an associated gain of 9.5 dB. Wide band amplifiers up to 40 GHz have already been proposed and some laboratories are developing components capable of working at 60 GHz or even higher.

In view of these advances, we considered that it would be useful for microwave designers and engineers to have a book which covered the appropriate theoretical background in order to allow the optimal usage of such active two-port elements. We hope also that the book will be of practical value and with this in mind have included examples of applications.

Chapter 1 takes as its starting point the concepts of current, voltage and impedance to introduce the scattering matrix, its basic properties and the precautions necessary when employing it at high frequencies. Chapter 2 draws upon this theoretical basis to define the concepts of gain, matching and stability. As the minimum noise factor is not necessarily equivalent to the maximum gain, we shall show how to obtain the best compromise which follows from the concept of minimum figure of merit. Chapter 3 gives three basic practical examples of amplifier design involving the application of microstrip technology. The last section introduces a novel generalization of the critical instability circle concept for use in designing maximum instability oscillators. Finally, measuring aspects are considered in Chapter 4, which is more specifically devoted to the network analyser and methods for the evaluation of noise.

I wish to express my thanks to Madame Dubois for typing the manuscript and to M. Morant for the drawings.

Chapter 1
Scattering matrix

1.1. Current and voltage scattering matrix

We shall begin by defining the **current and voltage reflection coefficients** of a one-port network (Fig. 1.1).

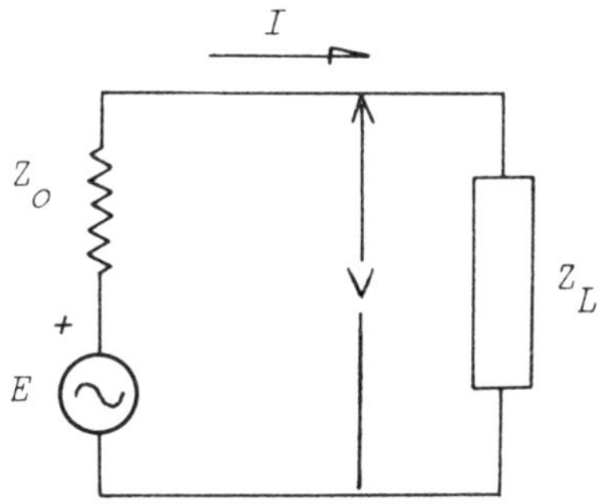

Fig. 1.1 *One-port network.*

A load Z_L is connected across the terminals of a voltage source E of internal impedance Z_0. The complex current and voltage are given respectively by

$$I = \frac{E}{Z_0 + Z_L} \tag{1.1}$$

and

$$V = \frac{EZ_L}{Z_0 + Z_L} \tag{1.2}$$

When the load impedance is equal to the conjugate of the internal impedance of the source, the system is said to be **power matched**:

$$Z_L = Z_0{}^*$$

(the asterisk indicates the conjugate).

The **incident current** is the current flowing under the matched condition:

$$I_i = \frac{E}{Z_0 + Z_0^*} = \frac{E}{2R_0} \tag{1.3}$$

where R_0 is the real part of Z_0.

The **incident voltage** is the voltage across the load under these conditions:

$$V_i = \frac{EZ_0^*}{Z_0 + Z_0^*} = \frac{EZ_0^*}{2R_0} \tag{1.4}$$

The reflected current and voltage components are then defined as follows:

$$I_r = I_i - I \tag{1.5}$$

$$V_r = V - V_i \tag{1.6}$$

Let us express the **reflected current** component in terms of the incident current:

$$I_r = \frac{E}{Z_0 + Z_0^*} - \frac{E}{Z_0 + Z_L} = \frac{E}{Z_0 + Z_0^*}\left(1 - \frac{Z_0 + Z_0^*}{Z_0 + Z_L}\right)$$

Hence

$$I_r = \frac{Z_L - Z_0^*}{Z_L + Z_0} I_i \tag{1.7}$$

Similarly, let us express the **reflected voltage** as a function of the incident voltage:

$$V_r = V_i\left(\frac{V}{V_i} - 1\right) = V_i\left(\frac{Z_L}{Z_0^*}\frac{Z_0 + Z_0^*}{Z_0 + Z_L} - 1\right)$$

or

$$V_r = \frac{Z_0}{Z_0^*}\frac{Z_L - Z_0^*}{Z_L + Z_0} V_i \tag{1.8}$$

The current and voltage reflection coefficients S_I and S_V are derived from relations (1.7) and (1.8) respectively:

$$S_I = \frac{Z_L - Z_0^*}{Z_L + Z_0} \tag{1.9}$$

$$S_V = \frac{Z_0}{Z_0^*}\frac{Z_L - Z_0^*}{Z_L + Z_0} \tag{1.10}$$

If the impedance Z_0 is real, $Z_0 = Z_0^* = R_0$; only in this case are the current and voltage reflection coefficients identical. Putting $S_I = S_V = S$ we have

$$S = \frac{Z_L - R_0}{Z_L + R_0} \tag{1.11}$$

The **normalized impedance** $z = Z/R_0$ is often used; hence the reflection coefficient may be written

$$S = \frac{z_L - 1}{z_L + 1} \tag{1.12}$$

and

$$z_L = \frac{1+S}{1-S} \tag{1.13}$$

The real part of the normalized load impedance is $\frac{1}{2}(z_L + z_L{}^*)$. Now

$$z_L + z_L{}^* = \frac{1+S}{1-S} + \frac{1+S^*}{1-S^*} = \frac{1+S-S^*-SS^*+1+S^*-S-SS^*}{(1-S)(1-S)^*}$$

and therefore

$$z_L + z_L{}^* = 2\frac{1-|S|^2}{|1-S|^2} \tag{1.14}$$

where $|S|^2 = SS^*$. From (1.12) we find

$$1-|S|^2 = 2\frac{z_L + z_L{}^*}{|1+z_L|^2} \tag{1.15}$$

which shows that $|S|$ is less than unity for a load with positive real part.

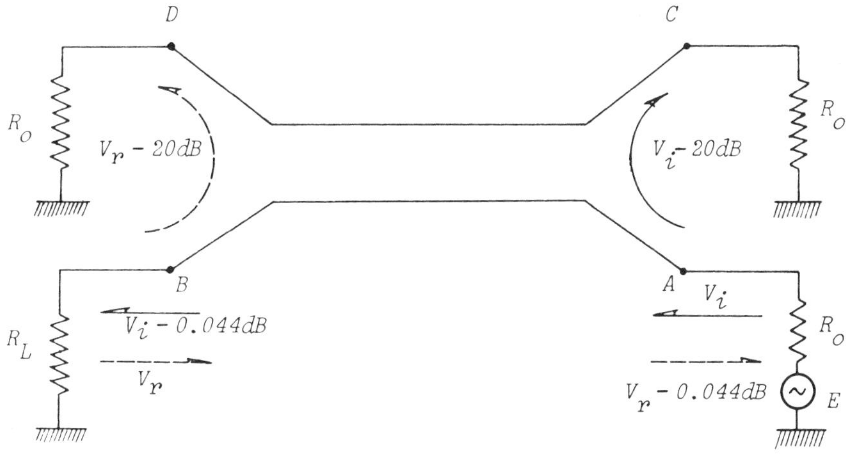

Fig. 1.2 Measurement of S_V using a directional coupler.

Note

Incident and reflected voltages can be measured directly, e.g. by using a **directional coupler** such as that shown in Fig. 1.2. Here we have a 20 dB coupler, which means that by connecting a generator at A we will find at B the fraction $(99/100)^{\frac{1}{2}} V_i$ (or in decibels V_i – 0.044 dB) while the fraction $V_i/10$ (or V_i – 20 dB) will be present at C. Assuming that the coupler is ideal, output D is perfectly isolated and therefore no voltage is detected when $R_L = R_0$. If the unknown load R_L is not equal to R_0, a reflected voltage appears: at A we find V_r – 0.044 dB and at D we find V_r – 20 dB. Finally, by measuring the voltages at C and D—in practice this is actually a power measurement —we directly obtain the ratio

$$\frac{V_r}{V_i} = \frac{R_L - R_0}{R_L + R_0} = S_V = S$$

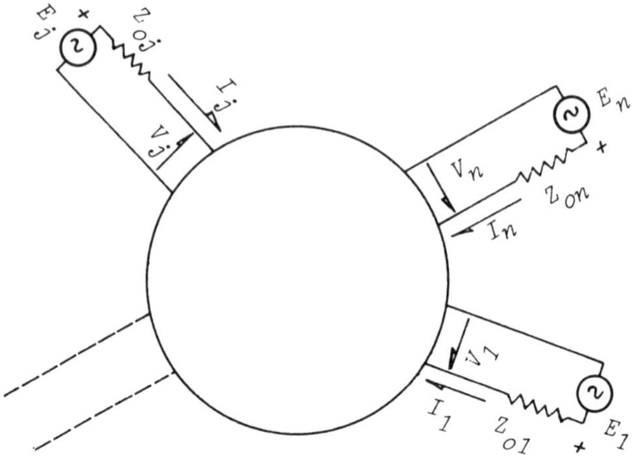

Fig. 1.3 *n-port network.*

We now consider the *n*-port network shown in Fig. 1.3. From definitions (1.3) and (1.4), we obtain

$$V_i = Z_0{}^* I_i \tag{1.16}$$

Moreover, combining (1.7) and (1.8) yields

$$V_r = \frac{Z_0}{Z_0{}^*} \frac{I_r}{I_i} V_i$$

and hence

$$V_r = Z_0 I_r \tag{1.17}$$

Generalizing (1.16) and (1.17) to the case of the *ı*-port network gives the matrix equations

$$[V_i] = [Z_0{}^*][I_i]$$

$$[V_r] = [Z_0][I_r]$$

where

$$[Z_0] = \begin{bmatrix} Z_{01} & & 0 \\ & \ddots\ Z_{0j}\ \ddots & \\ 0 & & Z_{0n} \end{bmatrix}$$

Similarly, (1.5) and (1.6) become

$$[I] = [I_i] - [I_r]$$
$$[V] = [V_i] + [V_r]$$

The impedance matrix $[Z]$ of the network is defined by

$$[V] = [Z][I]$$

Finally, (1.7) and (1.8) give

$$[I_r] = [S^I][I_i] \qquad (1.18)$$

$$[V_r] = [S^V][V_i] \qquad (1.19)$$

where $[S^I]$ and $[S^V]$ are the **current and voltage matrices** respectively.

We propose to express $[S^I]$ in terms of the impedance matrix $[Z]$ and $[S^V]$ in terms of the admittance matrix $[Y]$. Note that $[Z]$ and $[Y]$ are reciprocals of each other. Using the matrix equations introduced above, we can write

$$[V_r] = [V] - [V_i] = [Z][I] - [Z_0{}^*][I_i]$$

Now

$$[V_r] = [Z_0][I_r] \quad \text{and} \quad [I] = [I_i] - [I_r]$$

Hence

$$[Z_0][I_r] = [Z][I_i] - [Z][I_r] - [Z_0{}^*][I_i]$$

On rearranging, this becomes

$$([Z] + [Z_0])[I_r] = ([Z] - [Z_0{}^*])[I_i]$$

and so

$$[I_r] = ([Z] + [Z_0])^{-1}([Z] - [Z_0{}^*])[I_i]$$

Also

$$[I_r] = [S^I][I_i]$$

Thus

$$[S^I] = ([Z] + [Z_0])^{-1}([Z] - [Z_0{}^*]) \qquad (1.20)$$

Let us now express $[S^V]$ in terms of $[Y]$. We can start from

$$[I] = [Y][V]$$

and hence

$$[I_i] - [I_r] = [Y][V_i] + [Y][V_r]$$

Now from

$$[V_i] = [Z_0{}^*][I_i]$$

we obtain

$$[I_i] = [Y_0{}^*][V_i]$$

and similarly we have

$$[I_r] = [Y_0][V_r]$$

which gives

$$-([Y_o]+[Y])[V_r] = ([Y]-[Y_0{}^*])[V_i]$$

or

$$[V_r] = -([Y]+[Y_0])^{-1}([Y]-[Y_0{}^*])[V_i]$$

Equating with

$$[V_r] = [S^V][V_i]$$

gives

$$[S^V] = -([Y]+[Y_0])^{-1}([Y]-[Y_0{}^*]) \tag{1.21}$$

1.2. [*S*] matrix—link with the impedance and admittance matrices

We define a vector $[a]$, known as the **incident wave**, by the relation

$$[a] = \frac{([Z_0]+[Z_0{}^*])^{\frac{1}{2}}}{\sqrt{2}}[I_i] \tag{1.22}$$

where $[I_i] = [Z_0{}^*]^{-1}[V_i]$. As the real part of Z_{0j} is R_{0j}, the detailed notation is

$$\begin{bmatrix} a_1 \\ \vdots \\ a_j \\ \vdots \\ a_n \end{bmatrix} = \begin{bmatrix} (R_{01})^{\frac{1}{2}} & & 0 \\ & \ddots & \\ & (R_{0j})^{\frac{1}{2}} & \\ & & \ddots \\ 0 & & (R_{0n})^{\frac{1}{2}} \end{bmatrix} \times \begin{bmatrix} I_{i1} \\ \vdots \\ I_{ij} \\ \vdots \\ I_{in} \end{bmatrix}$$

We now define the **reflected wave** in terms of the vector $[b]$:

$$[b] = \frac{([Z_0]+[Z_0{}^*])^{\frac{1}{2}}}{\sqrt{2}}[I_r] \tag{1.23}$$

where

$$[I_r] = [Z_0]^{-1}[V_r]$$

From (1.22) and (1.23), we obtain

$$[I_i] = [R_0]^{-\frac{1}{2}}[a] \quad \text{and} \quad [I_r] = [R_0]^{-\frac{1}{2}}[b]$$

respectively.

The **generalized scattering matrix** $[S]$ is therefore defined by

$$[b] = [S][a] \tag{1.24}$$

Let us express $[S]$ in terms of $[S^I]$. From $[I_r] = [S^I][I_i]$ it follows that

$$[R_0]^{-\frac{1}{2}}[b] = [S^I][R_0]^{-\frac{1}{2}}[a]$$

or

$$[b] = [R_0]^{\frac{1}{2}}[S^I][R_0]^{-\frac{1}{2}}[a]$$

Hence, after equating with (1.24),

$$[S] = [R_0]^{\frac{1}{2}}[S^I][R_0]^{-\frac{1}{2}} \tag{1.25}$$

The term in the ith row and jth column is

$$S_{ij} = S^I{}_{ij}\frac{(R_{0i})^{\frac{1}{2}}}{(R_{0j})^{\frac{1}{2}}} \tag{1.26}$$

$[S]$ can be written directly in terms of $[S^V]$ as $[S^V]$ can be expressed simply in terms of $[S^I]$. In fact, from $[I_r] = [S^I][I_i]$ it now follows that

$$[V_r] = [Z_0][S^I][Z_0{}^*]^{-1}[V_i]$$

and consequently

$$[S^I] = [Z_0]^{-1}[S^V][Z_0{}^*] \tag{1.27}$$

or term by term

$$S^I{}_{ij} = S^V{}_{ij}\frac{Z_{0j}{}^*}{Z_{0i}} \tag{1.28}$$

In the case where the impedance of the generators is real, i.e. when $Z_{0j} = Z_{0j}{}^* = R_{0j}\,(\forall j \in 1, n)$, identity (1.28) can be rewritten in the form

$$S^I{}_{ij} = S^V{}_{ij}\frac{R_{0j}}{R_{0i}}$$

If, in addition, $R_{01} = \ldots = R_{0j} = \ldots = R_{0n}$, then we have

$$[S] = [S^I] = [S^V] \tag{1.29}$$

Returning to the vectors $[a]$ and $[b]$, we note that they have been defined in terms of the incident and reflected currents respectively. Moreover, the formulation in terms of the incident and reflected voltages was obvious. Let us now give a formulation introducing the vectors $[I]$ and $[V]$. For this purpose, consider the jth port shown in Fig. 1.4.

We have the equalities

$$E_j = V_j + Z_{0j}I_j$$
$$V_j = V_{ij} + V_{rj}$$

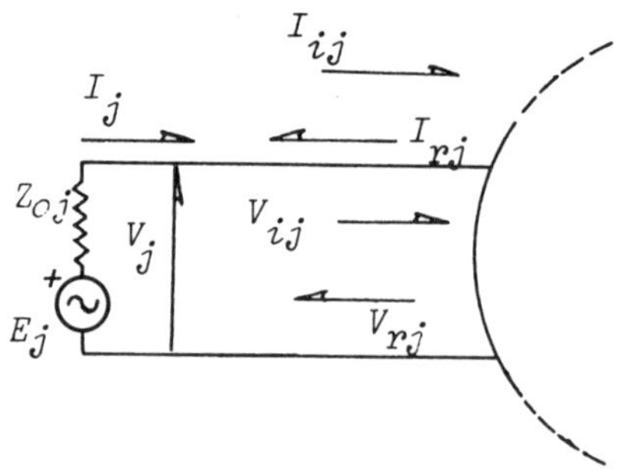

Fig. 1.4 *Currents and voltages at port j.*

Using $I_j = I_{ij} - I_{rj}$ and (1.16) and (1.17) we obtain

$$V_{ij} = Z_{0j}{}^* I_{ij}$$
$$V_{rj} = Z_{0j} I_{rj}$$

We therefore have

$$V_j + Z_{0j} I_j = Z_{0j}{}^* I_{ij} + Z_{0j} I_{rj} + Z_{0j} I_{ij} - Z_{0j} I_{rj}$$
$$V_j + Z_{0j} I_j = 2R_{0j} I_{ij}$$

Now, from (1.22)

$$a_j = R_{0j}{}^{\frac{1}{2}} I_{ij}$$

which implies

$$a_j = \frac{V_j + Z_{0j} I_j}{2R_{0j}{}^{\frac{1}{2}}} \tag{1.30}$$

where a_j is the jth term of the column matrix

$$[a] = \tfrac{1}{2}[R_0]^{-\frac{1}{2}}([V] + [Z_0][I]) \tag{1.31}$$

$V_j - Z_{0j}{}^* I_j$ can be expressed by

$$V_j - Z_{0j}{}^* I_j = Z_{0j}{}^* I_{ij} + Z_{0j} I_{rj} - Z_{0j}{}^* I_{ij} + Z_{0j}{}^* I_{rj}$$

or

$$V_j - Z_{0j}{}^* I_j = 2R_{0j} I_{rj}$$

From (1.23) $b_j = R_{0j}{}^{\frac{1}{2}} I_{rj}$, and hence

$$b_j = \frac{V_j - Z_{0j}{}^* I_j}{2R_{0j}{}^{\frac{1}{2}}} \tag{1.32}$$

and for the column matrix

$$[b] = \tfrac{1}{2}[R_0]^{-\frac{1}{2}}([V] - [Z_0{}^*][I]) \tag{1.33}$$

On adding (1.30) and (1.32), we obtain

$$a_j + b_j = \frac{V_j}{R_{0j}^{\frac{1}{2}}} + \frac{Z_{0j} - Z_{0j}^*}{2R_{0j}^{\frac{1}{2}}} I_j$$

In the special case when the impedance Z_{0j} is real, we have simply

$$a_j + b_j = \frac{V_j}{R_{0j}^{\frac{1}{2}}} \tag{1.34}$$

Now, subtracting (1.32) from (1.30) and making no assumptions concerning Z_{0j}, we have

$$a_j - b_j = R_{0j}^{\frac{1}{2}} I_j \tag{1.35}$$

The normalized voltage and current variables follow naturally from (1.34) and (1.35) and are expressed by

$$v_j = \frac{V_j}{R_{0j}^{\frac{1}{2}}} = a_j + b_j \tag{1.36}$$

and

$$i_j = R_{0j}^{\frac{1}{2}} I_j = a_j - b_j \tag{1.37}$$

respectively. From these new normalized variables, and therefore in the case where the impedance Z_{0j} is real ($Z_{0j} = R_{0j}$), we can define a **normalized impedance matrix** $[z]$ such that

$$[v] = [z][i]$$

Now

$$[v] = [R_0]^{-\frac{1}{2}}[V]$$

and

$$[i] = [R_0]^{\frac{1}{2}}[I]$$

with

$$[R_0] = \begin{bmatrix} R_{01} & & & & 0 \\ & \ddots & & & \\ & & R_{0j} & & \\ & & & \ddots & \\ 0 & & & & R_{0n} \end{bmatrix}$$

Hence

$$[V] = [R_0]^{\frac{1}{2}}[z][R_0]^{\frac{1}{2}}[I]$$

It therefore follows that

$$[Z] = [R_0]^{\frac{1}{2}}[z][R_0]^{\frac{1}{2}} \tag{1.38}$$

Similarly, the **normalized admittance matrix** is defined by

$$[i] = [y][v]$$

which gives

$$[Y] = [R_0]^{-\frac{1}{2}}[y][R_0]^{-\frac{1}{2}} \tag{1.39}$$

On applying (1.38) to a two-port network we obtain

$$[R_0]^{\frac{1}{2}} = \begin{bmatrix} R_{01}{}^{\frac{1}{2}} & 0 \\ 0 & R_{02}{}^{\frac{1}{2}} \end{bmatrix}$$

and this gives

$$[Z] = \begin{bmatrix} Z_{11} & Z_{12} \\ Z_{21} & Z_{22} \end{bmatrix} = \begin{bmatrix} z_{11}R_{01} & z_{12}(R_{01}R_{02})^{\frac{1}{2}} \\ z_{21}(R_{01}R_{02})^{\frac{1}{2}} & z_{22}R_{02} \end{bmatrix} \tag{1.40}$$

Conversely,

$$[z] = \begin{bmatrix} z_{11} & z_{12} \\ z_{21} & z_{22} \end{bmatrix} = \begin{bmatrix} Z_{11}/R_{01} & Z_{12}/(R_{01}R_{02})^{\frac{1}{2}} \\ Z_{21}/(R_{01}R_{02})^{\frac{1}{2}} & Z_{22}/R_{02} \end{bmatrix} \tag{1.41}$$

The scattering matrix $[S]$ can also be expressed in terms of the impedance and admittance matrices $[z]$ and $[y]$. From $[v] = [z][i]$ we obtain

$$([a]+[b]) = [z]([a]-[b])$$

or, on introducing the unit matrix $[I]$,

$$[I][a]+[I][b] = [z][a]-[z][b]$$

Hence

$$([z]+[I])[b] = ([z]-[I])[a]$$

and consequently

$$[S] = ([z]-[I])^{-1}([z]-[I]) \tag{1.42}$$

Similarly, we can calculate

$$[S] = -([y]-[I])^{-1}([y]-[I]) \tag{1.43}$$

Let us apply formula (1.42) to the case of a two-port network. Putting

$$\Delta z = (z_{11}+1)(z_{22}+1)-z_{12}z_{21}$$

we obtain

$$\begin{aligned} S_{11} &= \frac{(z_{22}+1)(z_{11}-1)-z_{12}z_{21}}{\Delta z} \\ S_{12} &= \frac{2z_{12}}{\Delta z} \\ S_{21} &= \frac{2z_{21}}{\Delta z} \end{aligned} \tag{1.44}$$

$$S_{22} = \frac{(z_{11}+1)(z_{22}-1)-z_{12}z_{21}}{\Delta z}$$

Apart from the sign, we can formally obtain the same expressions in terms of the y_{ij}:

$$S_{11} = \frac{(1-y_{11})(1+y_{22})+y_{12}y_{21}}{\Delta y}$$

$$S_{12} = \frac{-2y_{12}}{\Delta y}$$

$$S_{21} = \frac{-2y_{21}}{\Delta y} \qquad (1.45)$$

$$S_{22} = \frac{(1+y_{11})(1-y_{22})+y_{12}y_{21}}{\Delta y}$$

where

$$\Delta y = (1+y_{11})(1+y_{22})-y_{12}y_{21}$$

Finally, we still have to express the normalized matrices $[z]$ and $[y]$ in terms of $[S]$. Now, the matrix equation $[b] = [S][a]$ can be written as

$$([v]-[i]) = [S]([v]+[i])$$

From this we derive the two expressions

$$[z] = ([I]-[S])^{-1}([I]+[S]) \qquad (1.46)$$

and

$$[y] = ([I]+[S])^{-1}([I]-[S]) \qquad (1.47)$$

In the case of a two-port network, (1.46) gives

$$z_{11} = \frac{(1+S_{11})(1-S_{22})+S_{12}S_{21}}{\Delta S-}$$

$$z_{12} = \frac{2S_{12}}{\Delta S-}$$

$$z_{21} = \frac{2S_{21}}{\Delta S-} \qquad (1.48)$$

$$z_{22} = \frac{(1-S_{11})(1+S_{22})+S_{12}S_{21}}{\Delta S-}$$

$$\Delta S- = (1-S_{11})(1-S_{22})-S_{12}S_{21}$$

Moreover, relation (1.47) gives

$$
\begin{aligned}
y_{11} &= \frac{(1-S_{11})(1+S_{22})+S_{12}S_{21}}{\Delta S+} \\
y_{12} &= \frac{-2S_{12}}{\Delta S+} \\
y_{21} &= \frac{-2S_{21}}{\Delta S+} \\
y_{22} &= \frac{(1+S_{11})(1-S_{22})+S_{12}S_{21}}{\Delta S+} \\
\Delta S+ &= (1+S_{11})(1+S_{22})-S_{12}S_{21}
\end{aligned}
\tag{1.49}
$$

1.3. Scattering matrix in microwaves

At frequencies above a few hundred megahertz, the circuit dimensions may cease to be negligible compared with the wavelength. We recall the formula relating the wavelength λ_0 in free space to the frequency:

$$\lambda_0 = \frac{300}{F}$$

where λ_0 is in metres and F is in megahertz. It is therefore necessary to define **reference planes** at each port of the network (Fig. 1.5). In addition to time, voltage

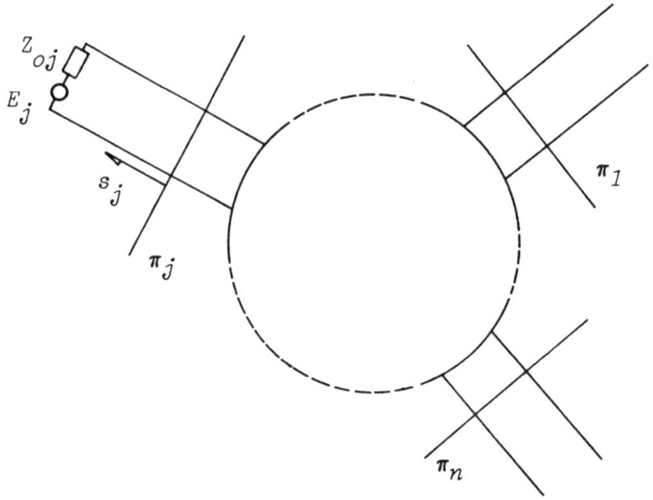

Fig. 1.5 *Reference planes.*

and current will also be functions of the distance s to these planes. We shall therefore denote by $v_j(s_j, t)$ and $i_j(s_j, t)$ the instantaneous voltages and currents measured (if measurement is possible!) at time t and distance s_j from port j. The problem is to know how $v_j(s_j, t)$ and $i_j(s_j, t)$ change along the line from the source to the wanted port at a given instant under given sinusoidal conditions. $V(s)$ and $I(s)$ are the corresponding complex quantities that we propose to determine.

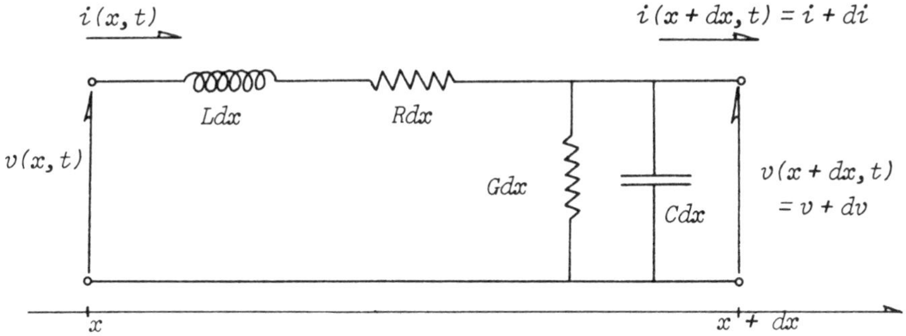

Fig. 1.6 Modelling of a transmission line.

A simple approach to the problem is to regard the source–network connections as transmission lines constituted by individual two-port networks connected in series (Fig. 1.6), each network being defined by its **linear constants**, i.e. the linear inductance $L\,(\mathrm{H\,m^{-1}})$, the linear capacitance $C\,(\mathrm{F\,m^{-1}})$, the linear conductance $G\,(\mathrm{S\,m^{-1}})$ and the linear resistance $R\,(\Omega\,\mathrm{m^{-1}})$, arranged as shown in Fig. 1.6. The equations take the form

$$v(x,t) - v(x+\mathrm{d}x, t) = -\mathrm{d}v = \left(Ri + L\frac{\partial i}{\partial t}\right)\mathrm{d}x$$

$$i(x,t) - i(x+\mathrm{d}x, t) = -\mathrm{d}i = \left(Gv + C\frac{\partial v}{\partial t}\right)\mathrm{d}x$$

or

$$-\frac{\partial V}{\partial x} = (R + \mathrm{j}L\omega)I$$

$$-\frac{\partial I}{\partial x} = (G + \mathrm{j}C\omega)V$$

Hence we obtain the wave equations

$$\frac{\partial^2 V}{\partial x^2} - \gamma^2 V = 0 \quad \text{and} \quad \frac{\partial^2 I}{\partial x^2} - \gamma^2 I = 0$$

where

$$\gamma = \alpha + \mathrm{j}\beta = \{(R + \mathrm{j}L\omega)(G + \mathrm{j}C\omega)\}^{\frac{1}{2}} \tag{1.50}$$

γ is the **propagation constant** whose real part α is the attenuation in nepers per metre and whose imaginary part β is the phase variation coefficient expressed in radians per metre.

In terms of γ, the solutions of the wave equations are

$$V = A\exp(-\gamma x) + B\exp(\gamma x) \quad \text{and} \quad I = \frac{1}{Z_c}\{A\exp(-\gamma x) - B\exp(\gamma x)\}$$

where

$$Z_c = \left(\frac{R + jL\omega}{G + jC\omega}\right)^{\frac{1}{2}} \tag{1.51}$$

Z_c is the **characteristic impedance** of the line. If $R = G = 0$, then $Z_c = (L/C)^{\frac{1}{2}}$, and from (1.50) $\alpha = 0$ and $\beta = \omega(LC)^{\frac{1}{2}}$.

Returning for a moment to the real domain, we obtain the following expression for the voltage:

$$v(x, t) = |A|\exp(-\alpha x)\cos(\omega t - \beta x + \varphi_A) + |B|\exp(\alpha x)\cos(\omega t + \beta x + \varphi_B)$$

which clearly shows that $v(x, t)$ is the sum of an incident wave propagating at **phase velocity** $v_\varphi = \omega/\beta$ and a reflected wave of phase velocity $-v_\varphi$. The same conclusions naturally apply to the current. Note that the wavelength λ_g on the line is related to the phase velocity by the equation $\lambda_g = v_\varphi/f$ or

$$\lambda_g = \frac{2\pi}{\omega}\frac{\omega}{\beta}$$

and therefore

$$\beta = \frac{2\pi}{\lambda_g}$$

Considering again the complex domain, and now as a function of the distance s to the reference plane, we obtain

$$V = V_0^+\exp(\gamma s) + V_0^-\exp(-\gamma s) \tag{1.52}$$

$$I = \frac{V_0^+}{Z_c}\exp(\gamma s) - \frac{V_0^-}{Z_c}\exp(-\gamma s) \tag{1.53}$$

The incident wave is characterized by

$$V_i = V_0^+\exp(\gamma s)$$

and

$$I_i = \frac{V_0^+}{Z_c}\exp(\gamma s)$$

The reflected wave is characterized by

$$V_r = V_0^-\exp(-\gamma s)$$

and

$$I_r = \frac{V_0^-}{Z_c} \exp(-\gamma s)$$

Hence

$$V = V_i + V_r \quad \text{and} \quad I = I_i - I_r$$

Returning to definitions (1.22) and (1.23) and taking as the generator impedance the impedance of the line connecting it to the jth port, we obtain

$$Z_{0j} + Z_{0j}^* = Z_{cj} + Z_{cj}^* = 2R_{0j}$$

which we rewrite in the form

$$a_j = R_{0j}^{\frac{1}{2}} I_{ij} = \frac{R_{0j}^{\frac{1}{2}}}{Z_{cj}^*} V_{ij}$$

$$b_j = R_{0j}^{\frac{1}{2}} I_{rj} = \frac{R_{0j}^{\frac{1}{2}}}{Z_{cj}} V_{rj}$$

In the case where the jth line considered is lossless, $\alpha_j = 0$ and $\gamma_j = \mathrm{j}\beta_j$. However, the characteristic impedance is real: $Z_{cj} = R_{0j}$. We therefore have

$$a_j = \frac{V_{0j}^+ \exp(\mathrm{j}\beta_j s_j)}{R_{0j}^{\frac{1}{2}}} \tag{1.54}$$

$$b_j = \frac{V_{0j}^- \exp(-\mathrm{j}\beta_j s_j)}{R_{0j}^{\frac{1}{2}}} \tag{1.55}$$

Still in the case of lossless lines, as source E_j is matched to the line of characteristic impedance R_{0j}, it can be shown that

$$a_j = \frac{V_j + R_{0j} I_j}{2R_{0j}^{\frac{1}{2}}} \quad \text{and} \quad b_j = \frac{V_j - R_{0j} I_j}{2R_{0j}^{\frac{1}{2}}}$$

Equations (1.54) and (1.55) clearly show that a_j is indeed an incident wave and that b_j is a reflected wave; this is illustrated in Fig. 1.7.

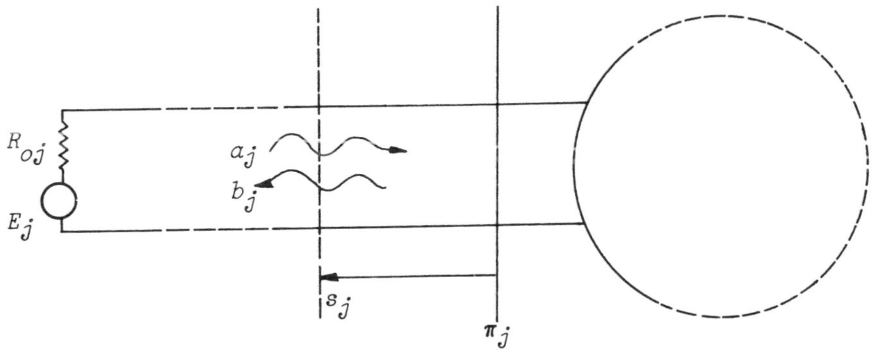

Fig. 1.7 Incident and reflected waves at a point on a line.

The scattering matrix, which we also redefine using the relation

$$[b] = [S][a]$$

has the same meaning as that defined above, but in this case $[S]$ depends on the reference plane π_j selected.

CHANGING REFERENCE PLANES

Let $[S_0]$ be the scattering matrix with reference planes π_{0j}. Let $[S]$ be the scattering matrix with new reference planes π_j separated from π_{0j} by the algebraic quantity s_j, as shown in Fig. 1.8.

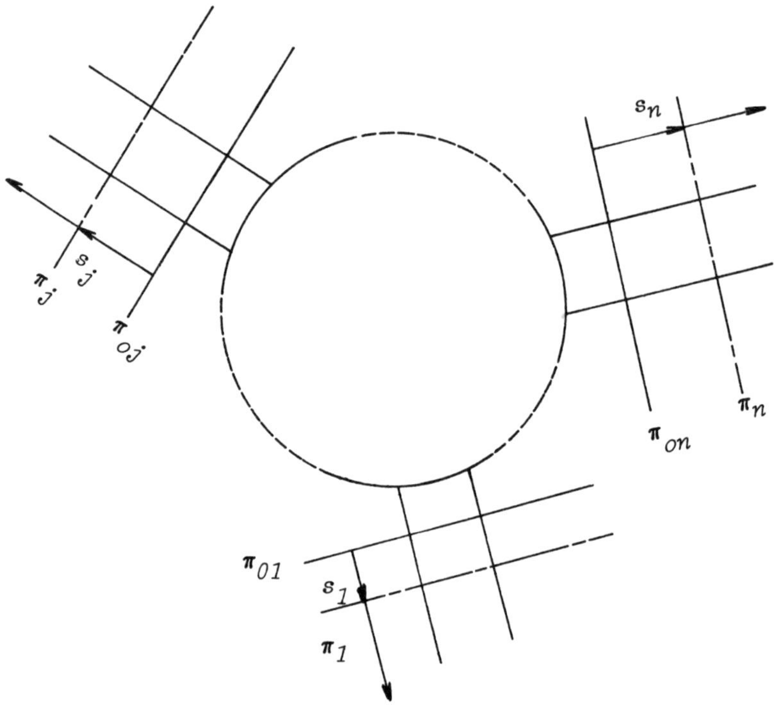

Fig. 1.8 *New reference planes.*

We have the relations

$$[b_0] = [S_0][a_0]$$

and

$$[b] = [S][a]$$

with

$$a_j = a_{0j}\exp(j\beta_j s_j) = a_{0j}\exp(j\varphi_j)$$
$$b_j = b_{0j}\exp(-j\beta_j s_j) = b_{0j}\exp(-j\varphi_j)$$

Let us introduce the $[\varphi]$ matrix:

$$[\varphi] = \begin{bmatrix} \exp(j\beta_1 s_1) & & & & 0 \\ & \ddots & & & \\ & & \exp(j\beta_j s_j) & & \\ & & & \ddots & \\ 0 & & & & \exp(j\beta_n s_n) \end{bmatrix}$$

This gives directly

$$[a] = [\varphi][a_0]$$
$$[b] = [\varphi]^{-1}[b_0]$$

Hence

$$[\varphi]^{-1}[b_0] = [S][\varphi][a_0]$$
$$[b_0] = [\varphi][S][\varphi][a_0]$$

or

$$[S_0] = [\varphi][S][\varphi] \quad (1.56)$$

Moreover,

$$[\varphi][b] = [S_0][\varphi]^{-1}[a]$$
$$[b] = [\varphi]^{-1}[S_0][\varphi]^{-1}[a]$$

and therefore

$$[S] = [\varphi]^{-1}[S_0][\varphi]^{-1} \quad (1.57)$$

What is the point of changing reference planes? By carefully selecting the new planes, we can obtain an $[S]$ matrix which is much simpler than the $[S_0]$ matrix. Moreover, in many cases the reference planes are dictated by measurement and we wish to use the reference planes of the item under test. In practice, when we measure the scattering parameters of a two-port network, we have the configuration shown in Fig. 1.9. The $[\varphi]$ matrix is given by

$$[\varphi] = \begin{bmatrix} \exp\left(-j2\pi\frac{d_1}{\lambda}\right) & 0 \\ \ddots & \\ 0 & \exp\left(-j2\pi\frac{d_2}{\lambda}\right) \end{bmatrix} = \begin{bmatrix} \exp(-j\varphi_1) & 0 \\ \ddots & \\ 0 & \exp(-j\varphi_2) \end{bmatrix}$$

Now

$$[S] = [\varphi]^{-1}[S_0][\varphi]^{-1}$$

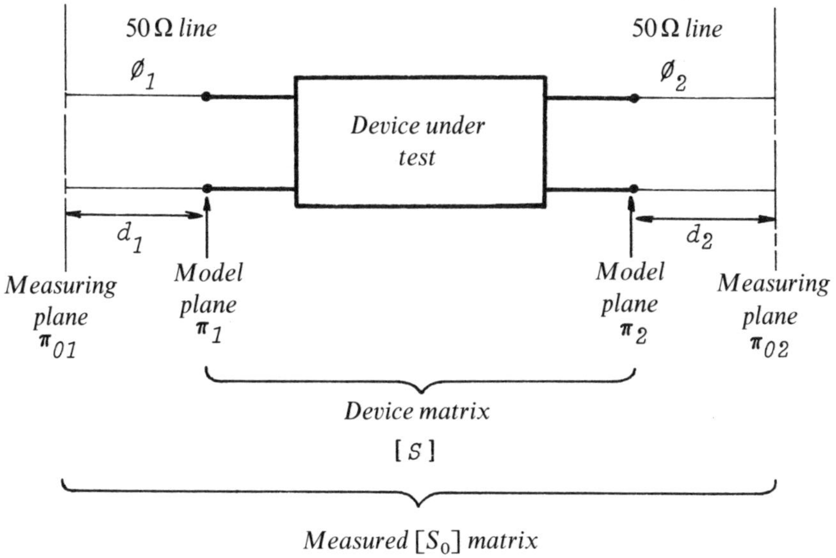

Fig. 1.9 *Transition from 'measuring' planes to 'model' planes.*

We therefore obtain

$$[S] = \begin{bmatrix} S^0{}_{11} \exp\left(j4\pi \dfrac{d_1}{\lambda}\right) & S^0{}_{12} \exp\left(j2\pi \dfrac{d_1 + d_2}{\lambda}\right) \\ S^0{}_{21} \exp\left(j2\pi \dfrac{d_1 + d_2}{\lambda}\right) & S^0{}_{22} \exp\left(j4\pi \dfrac{d_2}{\lambda}\right) \end{bmatrix} \tag{1.58}$$

1.4. Meaning and properties of the [S] matrix

To begin with, let us attempt to define the incident waves a_j and the reflected waves b_j in power terms. The **mean power** P_j transferred to port j (Fig. 1.10) is

$$P_j = \tfrac{1}{2}\operatorname{Re}\{V_j I_j{}^*\}$$

Using (1.34) and (1.35), we write

$$P_j = \tfrac{1}{2}\operatorname{Re}\{(a_j + b_j)(a_j{}^* - b_j{}^*)\}$$

when, on expansion, becomes

$$P_j = \tfrac{1}{2}\operatorname{Re}\{a_j a_j{}^* - a_j b_j{}^* + (a_j b_j{}^*)^* - b_j b_j{}^*\}$$

Now the quantity $-a_j b_j{}^* + (a_j b_j{}^*)^*$ is purely imaginary, and hence

$$P_j = \tfrac{1}{2}(|a_j|^2 - |b_j|^2) \tag{1.59}$$

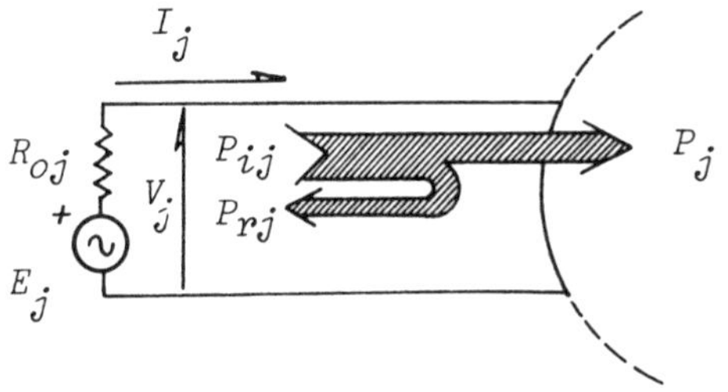

Fig. 1.10 *Power transfer to port j.*

Of course, we obtain the same result in the microwave domain since, in view of (1.54) and (1.55), Eqns (1.52) and (1.53) can be written

$$V_j = R_{0j}^{\frac{1}{2}}(a_j + b_j)$$

and

$$I_j = \frac{1}{R_{0j}^{\frac{1}{2}}}(a_j - b_j)$$

respectively. The quantity $\frac{1}{2}|a_j|^2$ represents the **incident power** P_{ij} and the quantity $\frac{1}{2}|b_j|^2$ represents the **reflected power** P_{rj}. If a passive and lossless n-port junction is assumed, conservation of energy implies the relation

$$\sum_{j=1}^{n} (|a_j|^2 - |b_j|^2) = 0$$

which in matrix notation becomes

$$[\tilde{a}]^*[a] = [\tilde{b}]^*[b]$$

where

$$[b] = [S][a]$$

and the tilde indicates matrix transposition. If the junction exhibits reciprocal behaviour, then the scattering matrix is symmetrical:

$$[S] = [\tilde{S}] \quad (S_{ij} = S_{ji})$$

Hence

$$[\tilde{b}]^* = \widetilde{[S]^*[a]^*} = [\tilde{a}]^*[\tilde{S}]^* = [\tilde{a}]^*[S]^*$$

and so

$$[\tilde{a}]^*[a] = [\tilde{a}]^*[S]^*[S][a]$$

Hence we obtain the following important result for a lossless matrix:

$$[S]^*[S] = [S][S]^* = [I] \tag{1.60}$$

or, explicitly,

$$\sum_k S_{ik} S_{kj}^* = \delta_{ij} \qquad \text{(Kronecker delta)} \tag{1.61}$$

By way of illustration, let us apply (1.60) to a reactive (i.e. passive and lossless) reciprocal two-port network. We obtain the four equalities

$$\begin{aligned} |S_{11}|^2 + |S_{12}|^2 &= 1 \quad \text{(a)} \\ S_{11}^* S_{12} + S_{12}^* S_{22} &= 0 \quad \text{(b)} \\ S_{12}^* S_{11} + S_{22}^* S_{12} &= 0 \quad \text{(c)} \\ |S_{12}|^2 + |S_{22}|^2 &= 1 \quad \text{(d)} \end{aligned} \qquad \text{(b)} \equiv \text{(c)}$$

The following relations between the moduli are derived from (a) and (d):

$$|S_{11}| = |S_{22}| = a \quad (0 \leqslant a \leqslant 1)$$

Hence

$$S_{11} = a \exp(\mathrm{j}\varphi_{11})$$

$$S_{22} = a \exp(\mathrm{j}\varphi_{22})$$

and

$$S_{12} = S_{21} = (1 - a^2)^{\frac{1}{2}} \exp(\mathrm{j}\theta)$$

($\varphi_{12} = \varphi_{21} = \theta$ since the two-port network is reciprocal). The phase relations

$$\exp\{\mathrm{j}(\theta - \varphi_{11})\} + \exp\{\mathrm{j}(\varphi_{22} - \theta)\} = 0$$

or

$$\theta - \varphi_{11} = \varphi_{22} - \theta \pm \pi$$

are derived from (b) or (c). Therefore

$$\theta = \frac{\varphi_{11} + \varphi_{22}}{2} \pm \frac{\pi}{2}$$

The scattering matrix of a passive, reciprocal and lossless two-port network is finally written in the form

$$[S] = \begin{bmatrix} a \exp(\mathrm{j}\varphi_{11}) & \pm \mathrm{j}(1 - a^2)^{\frac{1}{2}} \exp\left(\mathrm{j}\dfrac{\varphi_{11} + \varphi_{22}}{2}\right) \\ \pm \mathrm{j}(1 - a^2)^{\frac{1}{2}} \exp\left(\dfrac{\mathrm{j}\varphi_{11} + \varphi_{22}}{2}\right) & a \exp(\mathrm{j}\varphi_{22}) \end{bmatrix} \tag{1.62}$$

Let us now attempt to find the physical meaning of the S_{ij} terms of the scattering matrix. Although our analysis is in terms of a two-port network, this will in no way detract from the generality of the results.

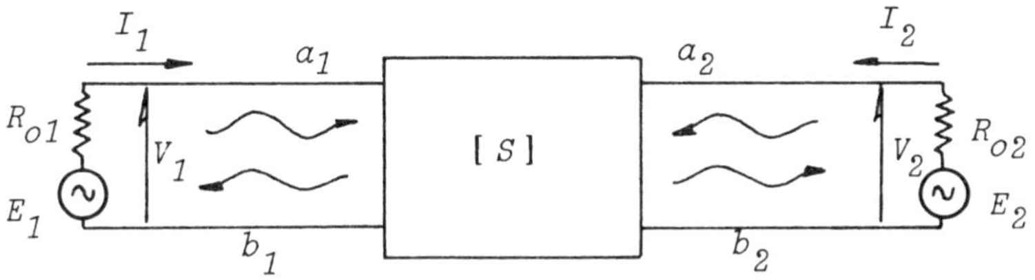

Fig. 1.11 Two-port network inserted between R_{01} and R_{02}.

Figure 1.11 restates the notation already used. We recall that

$$a_j = R_{0j}^{\frac{1}{2}} I_{ij} = \frac{V_j + R_{0j} I_j}{2R_{0j}^{\frac{1}{2}}} = \frac{E_j}{2R_{0j}^{\frac{1}{2}}} \qquad j = 1, 2$$

$$b_j = R_{0j}^{\frac{1}{2}} I_{rj} = \frac{V_j - R_{0j} I_j}{2R_{0j}^{\frac{1}{2}}} \qquad j = 1, 2$$

$$I_j = I_{ij} - I_{rj}$$

$$V_j = V_{ij} + V_{rj}$$

Meaning of S_{11}

$$S_{11} = \left.\frac{b_1}{a_1}\right|_{a_2 = 0}$$

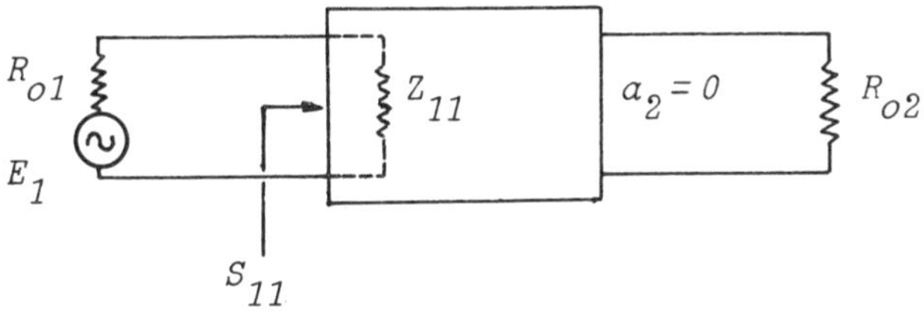

Fig. 1.12 Reflection coefficient S_{11}.

Now, $a_2 = 0$ implies that $E_2 = 0$, because then $V_2 = -R_{02} I_2$. This corresponds to the representation given in Fig. 1.12. However, we have the series of equations

$$S_{11} = \left.\frac{R_{01}^{\frac{1}{2}} I_{r1}}{R_{01}^{\frac{1}{2}} I_{i1}}\right|_{I_{i2} = 0} = \left.\frac{I_{r1}}{I_{i1}}\right|_{I_{i2} = 0} = \left.\frac{V_{r1}}{V_{i1}}\right|_{V_{i2} = 0}$$

The last equality follows from (1.7) and (1.8). Since $Z_L = Z_{11}$ and $Z_0 = R_{01}$ here, we have

$$S_{11} = \frac{Z_{11} - R_{01}}{Z_{11} + R_{01}} \tag{1.63}$$

S_{11} therefore represents the reflection coefficient at port 1 of the two-port network when port 2 is terminated on the reference resistance R_{02}. Another way of obtaining (1.63) is

$$S_{11} = \left.\frac{b_1}{a_1}\right|_{a_2=0} = \frac{V_1 - R_{01}I_1}{V_1 + R_{01}I_1} = \frac{Z_{11} - R_{01}}{Z_{11} + R_{01}}$$

since

$$Z_{11} = \left.\frac{V_1}{I_1}\right|_{E_2=0}$$

Let us now consider the square of the modulus of S_{11}:

$$|S_{11}|^2 = \left.\frac{|b_1|^2}{|a_1|^2}\right|_{a_2=0} = \left.\frac{\frac{1}{2}|b_1|^2}{\frac{1}{2}|a_1|^2}\right|_{a_2=0}$$

From relation (1.59), $\frac{1}{2}|b_1|^2$ represents the power reflected by port 1, also known as the **mismatch power**, which we shall denote here by P_1. The incident power satisfies

$$\frac{1}{2}|a_1|^2 = \frac{|E_1|^2}{8R_{01}} \tag{1.64}$$

where E_1 is the peak value. This is obviously the maximum power which the source can deliver and is the power effectively received by Z_{11} in the matched condition; denoting it by P_{A1} (the available power) we have

$$|S_{11}|^2 = \frac{P_1}{P_{A1}} \tag{1.65}$$

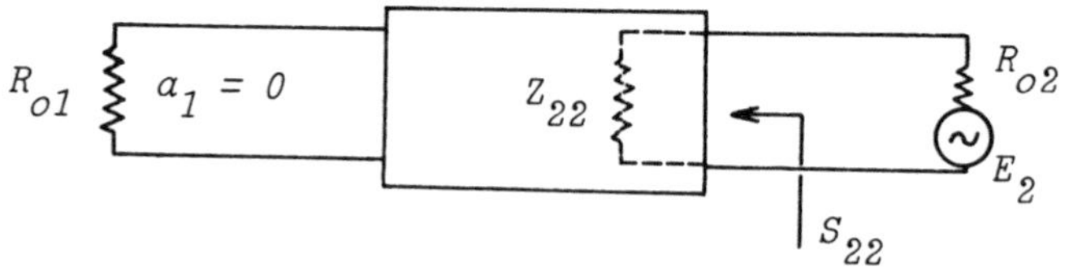

Fig. 1.13 Reflection coefficient S_{22}.

Meaning of S_{22}

$$S_{22} = \left.\frac{b_2}{a_2}\right|_{a_1=0}$$

In this case E_1 is zero. Figure 1.13 illustrates this configuration. The reflection coefficient aspect is given by

$$S_{22} = \frac{Z_{22} - R_{02}}{Z_{22} + R_{02}} \tag{1.66}$$

Since the reflected power is P_2 and the incident power is

$$P_{A2} = \frac{|E_2|^2}{8R_{02}}$$

the power S_{22} component is given by

$$|S_{22}|^2 = \frac{P_2}{P_{A2}} \tag{1.67}$$

Meaning of S_{21}

Once again, the source E_2 is zero:

$$S_{21} = \left.\frac{b_2}{a_1}\right|_{a_2=0}$$

This can be rewritten in terms of the incident and reflected currents as

$$S_{21} = \left.\frac{R_{02}{}^{\frac{1}{2}} I_{r2}}{R_{01}{}^{\frac{1}{2}} I_{i1}}\right|_{I_{i2}=0}$$

Now, $I_{i2} = 0$ implies that $I_2 = -I_{r2}$, and consequently

$$V_2 = -R_{02}I_2 = R_{02}I_{r2}$$

$$S_{21} = \frac{V_2/R_{02}{}^{\frac{1}{2}}}{\frac{1}{2}E_1/R_{01}{}^{\frac{1}{2}}} \tag{1.68}$$

The power term of S_{21} is very important. If we take the square of the modulus of S_{21}

$$|S_{21}|^2 = \left.\frac{\frac{1}{2}|b_2|^2}{\frac{1}{2}|a_1|^2}\right|_{a_2=0} = \frac{\frac{1}{2}V_2{}^2/R_{02}}{\frac{1}{8}E_1{}^2/R_{01}}$$

$\frac{1}{2}|b_2|^2$, when $a_2 = 0$, represents the power dissipated in the load R_{02}, which we shall denote by P_{L2}. Hence

$$|S_{21}|^2 = \frac{P_{L2}}{P_{A1}} \tag{1.69}$$

In the case of an active two-port network, $|S_{21}|^2$ is often called the **forward transducer power gain**, under given reference conditions. However, for a passive and lossless two-port network it is obvious that $P_{A1} = P_1 + P_{L2}$.

Meaning of S_{12}

$$S_{12} = \left.\frac{b_1}{a_2}\right|_{a_1=0}$$

This gives

$$S_{12} = \frac{V_1/R_{01}{}^{\frac{1}{2}}}{\frac{1}{2}E_2/R_{02}{}^{\frac{1}{2}}} \tag{1.70}$$

and in power terms

$$|S_{12}|^2 = \frac{P_{L1}}{P_{A2}} \tag{1.71}$$

The latter is often called the reverse transducer power gain.

Finally, let us examine the problem of connecting several two-port networks in cascade. In order to solve this problem, we often introduce the **wave matrix** T which is defined as shown in Fig. 1.14 or, in matrix format,

$$\begin{bmatrix} b_2 \\ a_2 \end{bmatrix} = [T]\begin{bmatrix} a_1 \\ b_1 \end{bmatrix} \tag{1.72}$$

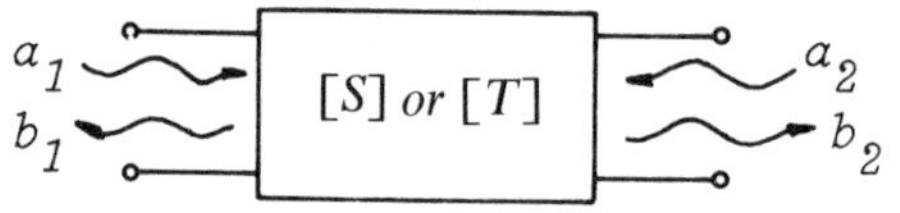

$$b_2 = T_{11}a_1 + T_{12}b_1 \qquad \text{(a)}$$

$$a_2 = T_{21}a_1 + T_{22}b_1 \qquad \text{(b)}$$

Fig. 1.14 *Representation of a two-port network by a wave matrix.*

Note that (a) and (b) (Fig. 1.14) enable us to write directly

$$b_1 = -\frac{T_{21}}{T_{22}}a_1 + \frac{1}{T_{22}}a_2$$

$$b_2 = \frac{T_{11}T_{22} - T_{12}T_{21}}{T_{22}}a_1 + \frac{T_{12}}{T_{22}}a_2$$

Hence

$$[S] = \begin{bmatrix} -\dfrac{T_{21}}{T_{22}} & \dfrac{1}{T_{22}} \\ \dfrac{T_{11}T_{22} - T_{12}T_{21}}{T_{22}} & \dfrac{T_{12}}{T_{22}} \end{bmatrix} \tag{1.73}$$

With the scattering matrix the reciprocal condition was expressed by $S_{12} = S_{21}$; using the wave matrix this condition is given by

$$T_{11}T_{22} - T_{12}T_{21} = 1 \tag{1.74}$$

The inverse identity of (1.73) is

$$[T] = \begin{bmatrix} \dfrac{S_{12}S_{21} - S_{11}S_{22}}{S_{12}} & \dfrac{S_{22}}{S_{12}} \\ -\dfrac{S_{11}}{S_{12}} & \dfrac{1}{S_{12}} \end{bmatrix} \tag{1.75}$$

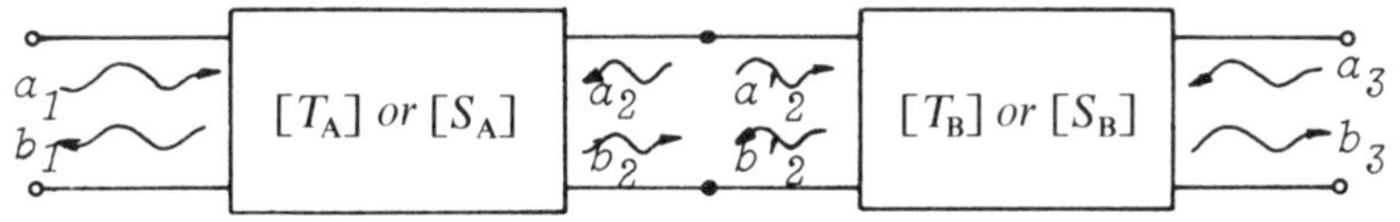

Fig. 1.15 *Cascading of two two-port networks.*

We propose to calculate the overall scattering matrix S_G of two cascaded two-port networks A and B with scattering matrices $[S_A]$ and $[S_B]$ and wave matrices $[T_A]$ and $[T_B]$ respectively; the incident and reflected waves have all been defined of course with respect to the same reference impedance R_0. Thus, for the network shown in Fig. 1.15 we can write

$$\begin{bmatrix} b_2 \\ a_2 \end{bmatrix} = \begin{bmatrix} a_2' \\ b_2' \end{bmatrix}$$

Now

$$\begin{bmatrix} b_2 \\ a_2 \end{bmatrix} = [T_A]\begin{bmatrix} a_1 \\ b_1 \end{bmatrix}$$

$$\begin{bmatrix} b_3 \\ a_3 \end{bmatrix} = [T_B]\begin{bmatrix} a_2' \\ b_2' \end{bmatrix}$$

and hence

$$\begin{bmatrix} b_3 \\ a_3 \end{bmatrix} = [T_B][T_A]\begin{bmatrix} a_1 \\ b_1 \end{bmatrix} = [T_G]\begin{bmatrix} a_1 \\ b_1 \end{bmatrix}$$

where

$$[T_G] = [T_B][T_A] \tag{1.76}$$

Taking the matrix product of $[T_B][T_A]$ in terms of the S parameters gives

$$T^G{}_{11} = \frac{(S^B{}_{12}S^B{}_{21} - S^B{}_{11}S^B{}_{22})(S^A{}_{12}S^A{}_{21} - S^A{}_{11}S^A{}_{22}) - S^A{}_{11}S^B{}_{22}}{S^A{}_{12}S^B{}_{12}}$$

$$T^G{}_{12} = \frac{(S^B{}_{12}S^B{}_{21} - S^B{}_{11}S^B{}_{22})S^A{}_{22} + S^B{}_{22}}{S^A{}_{12}S^B{}_{12}}$$

$$T^G{}_{21} = \frac{-(S^A{}_{12}S^A{}_{21} - S^A{}_{11}S^A{}_{22})S^B{}_{11} - S^A{}_{11}}{S^A{}_{12}S^B{}_{12}}$$

$$T^G{}_{22} = \frac{-S^B{}_{11}S^A{}_{22} + 1}{S^A{}_{12}S^B{}_{12}}$$

Substituting these expressions in (1.73) we have

$$\begin{aligned} S^{G}{}_{11} &= S^{A}{}_{11} + \frac{S^{A}{}_{12}S^{A}{}_{21}S^{B}{}_{11}}{D} \\ S^{G}{}_{12} &= \frac{S^{A}{}_{12}S^{B}{}_{12}}{D} \\ S^{G}{}_{21} &= \frac{S^{A}{}_{21}S^{B}{}_{21}}{D} \\ S^{G}{}_{22} &= S^{B}{}_{22} + \frac{S^{B}{}_{12}S^{B}{}_{21}S^{A}{}_{22}}{D} \end{aligned} \tag{1.77}$$

where

$$D = 1 - S^{A}{}_{22}S^{B}{}_{11}$$

We shall denote by $\circ$ the operator performing the product

$$[S_G] = [S_A] \circ [S_B] \tag{1.78}$$

in accordance with (1.77). If the $[S_A]$ and $[S_B]$ matrices are symmetrical, it is essential that the operator $\circ$ preserves this symmetry. Let us add in conclusion that the introduction of this operator finally enables us to dispense with the wave matrix; it gives us an indication of what to expect from cascading two-port networks.

Let us illustrate this for the case where the two two-ports represent two unilateral transistors, i.e. are such that

$$S^{A}{}_{12} = S^{B}{}_{12} = 0$$

In the limit we can write

$$[S_G] = \begin{bmatrix} S^{A}{}_{11} & 0 \\ \dfrac{S^{A}{}_{21}S^{B}{}_{21}}{1 - S^{A}{}_{22}S^{B}{}_{11}} & S^{B}{}_{22} \end{bmatrix} \tag{1.79}$$

Let us now assume that A alone is a non-reciprocal two-port network; we call the ratio

$$d_A = \frac{S^{A}{}_{12}}{S^{A}{}_{21}}$$

the non-reciprocity. It is easy to show, using (1.77), that connecting a reciprocal two-port B in cascade with A does not alter this ratio:

$$\frac{S^{A}{}_{12}}{S^{A}{}_{21}} = \frac{S^{G}{}_{12}}{S^{G}{}_{21}} = d_A$$

This result can of course be generalized to the case of a two-port network A inserted in a chain of reciprocal two-port networks.

Chapter 2

Two-port network amplifier theory

2.1. Transducer power gain of an active two-port network

The active two-port network is in fact the transistor whose S parameters measured with respect to a reference impedance R_0 are known. The transistor is driven by a generator of internal impedance Z_1 with a load impedance of Z_2. This can be represented by the diagram shown in Fig. 2.1.

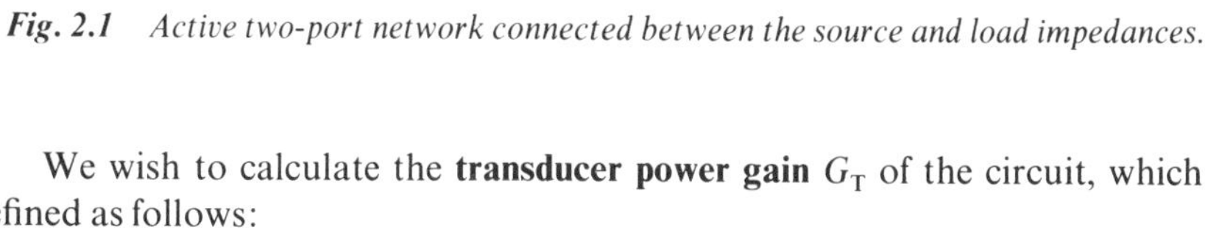

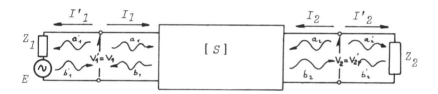

Fig. 2.1 *Active two-port network connected between the source and load impedances.*

We wish to calculate the **transducer power gain** G_T of the circuit, which is defined as follows:

$$G_T = \frac{\text{power dissipated in } Z_2}{\text{power available from source } (E, Z_1)} \tag{2.1}$$

If the two-port had been characterized by its $[S']$ matrix calculated by taking Z_1 and Z_2 as reference impedances, using (1.69) we would have obtained

$$G_T = |S'_{21}|^2$$

We propose to express the transducer power gain in terms of the S parameters and the reflection coefficients:

$$\Gamma_1 = \frac{Z_1 - R_0}{Z_1 + R_0}$$

at the input and

$$\Gamma_2 = \frac{Z_2 - R_0}{Z_2 + R_0}$$

at the output.

Let us define the relations between the incident and reflected waves more precisely.

Input side

$$a_1' = \frac{V_1' + R_0 I_1'}{2R_0^{\frac{1}{2}}} = \frac{V_1 - R_0 I_1}{2R_0^{\frac{1}{2}}} = b_1$$

$$b_1' = \frac{V_1' - R_0 I_1'}{2R_0^{\frac{1}{2}}} = \frac{V_1 + R_0 I_1}{2R_0^{\frac{1}{2}}} = a_1$$

Transforming the expression for b_1' gives

$$b_1' = \frac{V_1' - R_0 I_1'}{2R_0^{\frac{1}{2}}} \frac{Z_1 + R_0}{Z_1 + R_0}$$

$$b_1' = \frac{(V_1' + R_0 I_1' - 2R_0 I_1')(Z_1 - R_0 + 2R_0)}{2R_0^{\frac{1}{2}}(Z_1 + R_0)}$$

$$b_1' = \frac{(V_1' + R_0 I_1')(Z_1 - R_0) - 2R_0 Z_1 I_1' + 2R_0 V_1'}{2R_0^{\frac{1}{2}}(Z_1 + R_0)}$$

$$b_1' = \frac{V_1' + R_0 I_1'}{2R_0^{\frac{1}{2}}} \frac{Z_1 - R_0}{Z_1 + R_0} + \frac{2R_0(V_1' - Z_1 I_1')}{2R_0^{\frac{1}{2}}(Z_1 + R_0)}$$

Hence finally we obtain

$$b_1' = \Gamma_1 a_1' + \frac{R_0^{\frac{1}{2}}}{Z_1 + R_0} E$$

or again

$$a_1 = \Gamma_1 b_1 + \frac{R_0^{\frac{1}{2}}}{Z_1 + R_0} E \tag{2.2}$$

Output side

We have the identities

$$a_2' = b_2$$

$$b_2' = a_2$$

and of course

$$\frac{b_2'}{a_2'} = \frac{V_2' - R_0 I_2'}{V_2' + R_0 I_2'} = \Gamma_2$$

or equally

$$a_2 = \Gamma_2 b_2 \tag{2.3}$$

Returning to definition (2.1), we obtain

$$G_T = \frac{\frac{1}{2}(|a_2'|^2 - |b_2'|^2)}{\frac{1}{2}|E|^2/4Re\{Z_1\}}$$

The numerator can be written in the form

$$|a_2'|^2 - |b_2'|^2 = |b_2|^2 - |a_2|^2 = |b_2|^2(1 - |\Gamma_2|^2)$$

The denominator can be written as

$$\frac{|E|^2}{4Re\{Z_1\}} = |E|^2 \frac{1 - |\Gamma_1|^2}{2(Z_1 + Z_1^*)} \frac{1}{1 - |\Gamma_1|^2}$$

with (see Eqn (1.15))

$$1 - |\Gamma_1|^2 = 1 - \Gamma_1 \Gamma_1^* = 2R_0 \frac{Z_1 + Z_1^*}{|Z_1 + R_0|^2}$$

Hence

$$\frac{|E|^2}{4Re\{Z_1\}} = |E|^2 \frac{R_0}{|Z_1 + R_0|^2} \frac{1}{1 - |\Gamma_1|^2}$$

and therefore

$$G_T = \left|\frac{b_2}{E}\right|^2 \frac{|Z_1 + R_0|^2}{R_0} (1 - |\Gamma_1|^2)(1 - |\Gamma_2|^2) \tag{2.4}$$

We must now compute the ratio b_2/E; to do this, we must solve the system

$$b_1 = S_{11} a_1 + S_{12} a_2 \qquad \text{(a)}$$

$$b_2 = S_{21} a_1 + S_{22} a_2 \qquad \text{(b)}$$

$$a_1 = \Gamma_1 b_1 + \frac{R_0^{\frac{1}{2}}}{Z_1 + R_0} E \qquad \text{(c)}$$

$$a_2 = \Gamma_2 b_2 \qquad \text{(d)}$$

Equation (c) gives

$$\Gamma_1 b_1 = a_1 - \frac{R_0^{\frac{1}{2}}}{Z_1 + R_0} E$$

or in (a)

$$a_1 - \frac{R_0^{\frac{1}{2}}}{Z_1 + R_0} E = \Gamma_1 S_{11} a_1 + \Gamma_1 \Gamma_2 S_{12} b_2$$

and hence

$$a_1(1-\Gamma_1 S_{11}) = \frac{R_0^{\frac{1}{2}}}{Z_1+R_0} E + \Gamma_1 \Gamma_2 S_{12} b_2$$

It merely remains to introduce this last expression in (b):

$$(1-\Gamma_1 S_{11}) b_2 = \frac{R_0^{\frac{1}{2}}}{Z_1+R_0} E S_{21} + \Gamma_1 \Gamma_2 S_{12} S_{21} b_2 + S_{22} \Gamma_2 (1-\Gamma_1 S_{11}) b_2$$

which on rearranging gives

$$(1-\Gamma_1 S_{11})(1-\Gamma_2 S_{22}) b_2 - \Gamma_1 \Gamma_2 S_{12} S_{21} b_2 = \frac{R_0^{\frac{1}{2}}}{Z_1+R_0} S_{21} E$$

Finally, the expression for the ratio b_2/E is

$$\frac{b_2}{E} = \frac{R_0^{\frac{1}{2}}}{Z_1+R_0} \frac{S_{21}}{(1-\Gamma_1 S_{11})(1-\Gamma_2 S_{22}) - \Gamma_1 \Gamma_2 S_{12} S_{21}} \tag{2.5}$$

Using (2.4) and taking account of (2.5), we therefore obtain the required expression for the transducer power gain:

$$G_T = \frac{|S_{21}|^2 (1-|\Gamma_1|^2)(1-|\Gamma_2|^2)}{|(1-\Gamma_1 S_{11})(1-\Gamma_2 S_{22}) - \Gamma_1 \Gamma_2 S_{12} S_{21}|^2} \tag{2.6}$$

We can show that, if $Z_1 = Z_2 = R_0$,

$$\Gamma_1 = \Gamma_2 = 0$$

and therefore

$$G_T = |S_{21}|^2$$

Note 1

Equation (2.5) could also have been obtained by analytical reduction of the flowchart shown in Fig. 2.2. Using Mason's rule, we can easily solve a graph of this kind. The method is explained in Appendix A, in which the calculation of the voltage gain V_2/V_1 is taken as an example.

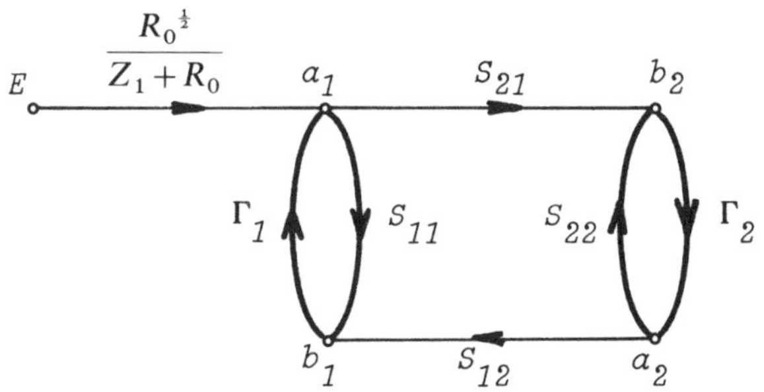

Fig. 2.2 Flowchart of a two-port network driven by a Thévenin generator.

Note 2

Let us return to the fact that $G_T = |S'_{21}|^2$. We saw that the active two-port network was in practice only characterized by its S parameters measured with respect to a single reference R_0 (namely $R_0 = 50\,\Omega$). Using $[S]$, we would like to find the matrix equation giving us the generalized scattering matrix $[S']$ expressed in terms of the reference impedances Z_i. The calculations are very lengthy; a typical demonstration will be found in an Appendix to *High Frequency Amplifiers* by R. S. Carson (see References). We shall give the result for the general case of an n-port network terminated in impedances Z_i. We have

$$[S'] = [A]^{-1}([S]-[\Gamma^*])([I]-[\Gamma][S])^{-1}[A^*] \tag{2.7}$$

where $[A]$ and $[\Gamma]$ are diagonal matrices. The term i, i of the $[\Gamma]$ matrix is given by

$$\Gamma_i = \frac{Z_i - R_0}{Z_i + R_0}$$

The term i, i of the $[A]$ matrix is given by

$$A_i = (1-\Gamma_i^*)\frac{(1-|\Gamma_i|^2)^{\frac{1}{2}}}{|1-\Gamma_i|} \tag{2.8}$$

On limiting (2.7) to the case of a two-port network, we obtain

$$\begin{aligned}
S'_{11} &= \frac{A_1^*}{A_1}\,\frac{(1-\Gamma_2 S_{22})(S_{11}-\Gamma_1^*)+\Gamma_2 S_{12}S_{21}}{D} \\
S'_{12} &= \frac{A_2^*}{A_1}\,\frac{S_{12}(1-|\Gamma_1|^2)}{D} \\
S'_{21} &= \frac{A_1^*}{A_2}\,\frac{S_{21}(1-|\Gamma_2|^2)}{D} \\
S'_{22} &= \frac{A_2^*}{A_2}\,\frac{(1-\Gamma_1 S_{11})(S_{22}-\Gamma_2^*)+\Gamma_1 S_{12}S_{21}}{D}
\end{aligned} \tag{2.9}$$

$$D = (1-\Gamma_1 S_{11})(1-\Gamma_2 S_{22}) - \Gamma_1\Gamma_2 S_{12}S_{21}$$

and it can easily be shown that $|S'_{21}|^2 = G_T$ (Eqn (2.6)).

Note 3

Let us give three alternative expressions for G_T. Expanding the denominator of (2.6) we obtain

$$G_T = \frac{|S_{21}|^2(1-|\Gamma_1|^2)(1-|\Gamma_2|^2)}{|1-\Gamma_1 S_{11}-\Gamma_2 S_{22}+\Delta\Gamma_1\Gamma_2|^2}$$

where ($\Delta = S_{11}S_{22}-S_{12}S_{21}$). Putting

$$S'_{11} = S_{11} + \frac{S_{12}S_{21}\Gamma_2}{1-\Gamma_2 S_{22}}$$

gives

$$1-\Gamma_1 S_{11} = 1-\Gamma_1 S'_{11}+\frac{\Gamma_1\Gamma_2 S_{12}S_{21}}{1-\Gamma_2 S_{22}}$$

and hence

$$G_T = \frac{|S_{21}|^2(1-|\Gamma_1|^2)(1-|\Gamma_2|^2)}{|1-\Gamma_1 S'_{11}|^2|1-\Gamma_2 S_{22}|^2}$$

Now, putting

$$S'_{22} = S_{22}+\frac{S_{12}S_{21}\Gamma_1}{1-\Gamma_1 S_{11}}$$

gives

$$G_T = \frac{|S_{21}|^2(1-|\Gamma_1|^2)(1-|\Gamma_2|^2)}{|1-\Gamma_1 S_{11}|^2|1-\Gamma_2 S'_{22}|^2}$$

2.2. Amplifier matching and stability

We want the source to deliver maximum power to the active two-port network and we also want the load to receive maximum power; this corresponds to the maximum transducer power gain $G_{T(max)}$.

It is therefore necessary that the input impedance of the two-port network with load impedance Z_2 should be equal to Z_1^* and that the output impedance of the two-port network with load impedance Z_1 (source E deactivated) should be equal to Z_2^*. Expressing this in terms of the reflection coefficient gives

$$\frac{b_1}{a_1} = \frac{Z_1^* - R_0}{Z_1^* + R_0} = \Gamma_1^*$$

Now, from the system

$$b_1 = S_{11}a_1 + S_{12}a_2$$

$$b_2 = S_{21}a_1 + S_{22}a_2$$

$$a_2 = \Gamma_2 b_2$$

(see Eqn (2.3)) we obtain

$$\frac{b_1}{a_1} = S_{11} + S_{12}\frac{a_2}{a_1}$$

with

$$\left(\frac{1}{\Gamma_2} - S_{22}\right)a_2 = S_{21}a_1$$

Hence

$$\Gamma_1^* = S_{11} + \frac{S_{12}S_{21}}{1/\Gamma_2 - S_{22}} \tag{2.10}$$

Similarly, we obtain

$$\Gamma_2^*(E = 0) = S_{22} + \frac{S_{12}S_{21}}{1/\Gamma_1 - S_{11}} \tag{2.11}$$

at the output side.

What we know in practice are the S parameters of the transistor selected at the required operating frequency. Therefore matching consists of determining Z_1 and Z_2, i.e. Γ_1 and Γ_2, in terms of the scattering parameters.

We now calculate Γ_1:

$$\Gamma_1 = S_{11}^* + \frac{S_{12}^* S_{21}^*}{1/\Gamma_2^* - S_{22}^*}$$

Now

$$\Gamma_2^* = \frac{S_{22} - \Delta\Gamma_1}{1 - S_{11}\Gamma_1}$$

where

$$\Delta = S_{11}S_{22} - S_{12}S_{21}$$

We substitute this expression for Γ_2^* in the relation for Γ_1:

$$\Gamma_1 = S_{11}^* + \frac{S_{12}^* S_{21}^*(S_{22} - \Delta\Gamma_1)}{1 - S_{11}\Gamma_1 - |S_{22}|^2 + S_{22}^*\Delta\Gamma_1}$$

On expansion this becomes

$$\begin{aligned}\Gamma_1(1-|S_{22}|^2) + \Gamma_1{}^2(\Delta S_{22}^* - S_{11}) \\ = \Gamma_1(\Delta S_{11}^* S_{22}^* - |S_{11}|^2 - \Delta S_{12}^* S_{21}^*) + S_{11}^* - S_{11}^*|S_{22}|^2 + S_{12}^* S_{21}^* S_{22}\end{aligned}$$

Hence, by regrouping and taking account of the fact that

$$\Delta S_{11}^* S_{22}^* - \Delta S_{12}^* S_{21}^* = |\Delta|^2$$

we obtain the following quadratic equation:

$$\begin{aligned}(S_{11} - \Delta S_{22}^*)\Gamma_1{}^2 + (|\Delta|^2 - |S_{11}|^2 + |S_{22}|^2 - 1)\Gamma_1 \\ + (S_{11}^* - S_{11}^*|S_{22}|^2 + S_{12}^* S_{21}^* S_{22}) = 0\end{aligned}$$

Note that the constant term is equal to C_1^* such that

$$C_1^* = S_{11}^* - \Delta^* S_{22} = (S_{11} - \Delta S_{22}^*)^*$$

and thus

$$\Gamma_1 = \frac{B_1 \pm (B_1{}^2 - 4|C_1|^2)^{\frac{1}{2}}}{2C_1}$$

$$B_1 = 1+|S_{11}|^2-|S_{22}|^2-|\Delta|^2$$
$$C_1 = S_{11}-\Delta S_{22}^* \tag{2.12}$$
$$\Delta = S_{11}S_{22}-S_{12}S_{21}$$

We now calculate Γ_2, and in similar fashion we obtain

$$\Gamma_2 = \frac{B_2 \pm (B_2{}^2-4|C_2|^2)^{\frac{1}{2}}}{2C_2}$$
$$B_2 = 1+|S_{22}|^2-|S_{11}|^2-|\Delta|^2 \tag{2.13}$$
$$C_2 = S_{22}-\Delta S_{11}^*$$

We now calculate $B_1{}^2-4|C_1|^2$. By expansion we obtain

$$B_1{}^2 = 1+2|S_{11}|^2-2|S_{22}|^2-2|\Delta|^2+|S_{11}|^4+|S_{22}|^4+|\Delta|^4-2|S_{11}|^2|S_{22}|^2 -2|\Delta|^2|S_{11}|^2+2|\Delta|^2|S_{22}|^2$$

$$|C_1|^2 = |S_{11}|^2-\Delta^* S_{11}S_{22}-\Delta S_{11}^* S_{22}^* +|\Delta|^2|S_{22}|^2$$
$$= |S_{11}|^2-2|S_{11}|^2|S_{22}|^2+|\Delta|^2|S_{22}|^2+S_{12}^* S_{21}^* S_{11}S_{22}+S_{12}S_{21}S_{11}^* S_{22}^*$$

Now, by taking the product $|\Delta|^2 = \Delta\Delta^*$ we have

$$S_{12}^* S_{21}^* S_{11}S_{22}+S_{12}S_{21}S_{11}^* S_{22}^* = |S_{11}|^2|S_{22}|^2+|S_{12}|^2|S_{21}|^2-|\Delta|^2$$

Hence

$$|C_1|^2 = |S_{11}|^2-|S_{11}|^2|S_{22}|^2+|S_{12}|^2|S_{21}|^2-|\Delta|^2+|\Delta|^2|S_{22}|^2$$

and therefore

$$B_1{}^2-4|C_1|^2 = 1-2|S_{11}|^2-2|S_{22}|^2+2|\Delta|^2+|S_{11}|^4+|S_{22}|^4+|\Delta|^4 +2|S_{11}|^2|S_{22}|^2-2|\Delta|^2|S_{11}|^2-2|\Delta|^2|S_{22}|^2-4|S_{12}|^2|S_{21}|^2$$

or

$$B_1{}^2-4|C_1|^2 = (1+|\Delta|^2-|S_{11}|^2-|S_{22}|^2)^2-4|S_{12}|^2|S_{21}|^2$$

This expression can finally be written in the form

$$B_1{}^2-4|C_1|^2 = 4(K^2-1)|S_{12}S_{21}|^2 \tag{2.14a}$$

with

$$K = \frac{1+|\Delta|^2-|S_{11}|^2-|S_{22}|^2}{2|S_{12}||S_{21}|}$$

The right-hand side of Eqn (2.14a) is invariant when subscripts 1 and 2 are interchanged and hence

$$B_1{}^2-4|C_1|^2 = B_2{}^2-4|C_2|^2 \tag{2.14b}$$

Let us now discuss the possibility of simultaneous input–output matching.

The discussion will be in terms of the values of the parameter K introduced above (Eqn (2.14a)) which is a characteristic of the two-port network and is even an invariant when the transistor is connected in series with reactive two-port networks (the demonstration is based on Eqn (1.77)).

To begin with, let us recall that impedances Z_1 and Z_2 have a real positive part and that consequently we have (see Eqns (1.14) or (1.15))

$$|\Gamma_1|<1 \quad \text{and} \quad |\Gamma_2|<1$$

If, however, the reflection coefficients presented to the input (Eqn (2.10)) and the output (Eqn (2.11)) have a modulus greater than unity, they correspond to impedances with a negative real part and as a result the circuit is unstable because it is a potential oscillator generator.

Case when $|K| = 1$

In this case $B_1^2 = 4|C_1|^2$ and $B_2^2 = 4|C_2|^2$ and hence $|\Gamma_1| = |\Gamma_2| = 1$; in practice the amplifier cannot be matched.

Case for $|K| > 1$

(a) $K > 1$. Let us state the two solutions of Eqn (2.12):

$$\Gamma_1' = \frac{B_1-(B_1^2-4|C_1|^2)^{\frac{1}{2}}}{2C_1}$$

$$\Gamma_1'' = \frac{B_1+(B_1^2-4|C_1|^2)^{\frac{1}{2}}}{2C_1}$$

We also have Γ_2' and Γ_2'' from (2.13). Taking the product of the moduli, we obtain the important result

$$\begin{aligned} |\Gamma_1'||\Gamma_1''| &= 1 \\ |\Gamma_2'||\Gamma_2''| &= 1 \end{aligned} \tag{2.15}$$

It therefore follows that input and output can be matched simultaneously since one of the two solutions (2.12) will have a modulus less than unity and will correspond to the solution of (2.13) whose modulus is also less than unity. To be precise, if $|\Delta|$ is less than unity the amplifier is unconditionally stable, i.e. it is stable irrespective of the values of the source and load impedances Z_1 and Z_2. The pair of solutions to be adopted for simultaneous matching is Γ_1', Γ_2' as in this case we have simultaneously $B_1>0$, $B_2>0$. If $|\Delta|$ is greater than unity, stability is conditional but the amplifier can still be matched; this time the pair of solutions to be adopted is Γ_1'', Γ_2'' because $B_1<0$ and $B_2<0$.

Therefore, when K is greater than unity input and output matching is always possible. If we then introduce the pairs $\{\Gamma_1', \Gamma_2'\}$ or $\{\Gamma_1'', \Gamma_2''\}$, depending on the sign of B_1 in the expression for the transducer power gain G_T (Eqn (2.6)), we obtain in the former case an effective maximum gain $G_{T(max)}$ whereas in the latter case we have a minimum $G_{T(min)}$ which, conversely, ensures optimum conditional stability. $G_{T(min)}$ ($G_{T(max)}$) can be expressed simply in terms of K and the non-reciprocity:

$$G_{T(min)}(G_{T(max)}) = \frac{|S_{21}|}{|S_{12}|}|K \pm (K^2-1)^{\frac{1}{2}}| \tag{2.16}$$

with the negative sign for $B_1 > 0$ (max) and the positive sign for $B_1 < 0$ (min).

This direct calculation of $G_{T(max)}$ is somewhat tedious, whereas we quickly arrive at the result by using the invariance properties of K and the ratio S_{21}/S_{12} (see Chapter 1, end of Section 4). We reproduce below the demonstration given by P. Grivet in *Microwave Circuits and Amplifiers*, Vol. 2, Fascicle 2 (see References).

For a matched amplifier $S^A{}_{11} = S^A{}_{22} = 0$, and consequently

$$|\Delta^A| = |S^A{}_{12}S^A{}_{21}|$$

Hence

$$K = \frac{1+|\Delta^A|^2}{2|\Delta^A|}$$

and therefore

$$|S^A{}_{12}||S^A{}_{21}| = K \pm (K^2-1)^{\frac{1}{2}}$$

As unconditional stability requires $|\Delta^A| < 1$, only the minus sign has to be considered. Moreover

$$\frac{|S^A{}_{21}|}{|S^A{}_{12}|} = \frac{|S_{21}|}{|S_{12}|}$$

The term-by-term product of these last two equalities gives

$$G_{T(max)} = |S^A{}_{21}|^2 = \frac{|S_{21}|}{|S_{12}|}\{K-(K^2-1)^{\frac{1}{2}}\} \tag{2.17}$$

Now, if $|\Delta^A|$ is greater than unity it is under matching conditions so that the gain G_T is minimum:

$$G_{T(min)} = \left|\frac{S_{21}}{S_{12}}\right|\{K+(K^2-1)^{\frac{1}{2}}\}$$

(b) $K < -1$. We can show that in this case it will be necessary to select either the pair (Γ_1', Γ_2'') or the pair (Γ_1'', Γ_2'), i.e. to associate a root of modulus greater than unity with a root of modulus less than unity. The transistor is therefore not simultaneously matchable; moreover, it is naturally unstable and therefore cannot be used as an amplifier.

Case for $|K| < 1$

In this case $B_1{}^2 - 4|C_1|^2 < 0$ and similarly $B_2{}^2 - 4|C_2|^2 < 0$. Hence

$$|\Gamma_i'| = |\Gamma_i''| = 1 \qquad i = 1, 2$$

Matching is impossible, and the amplifier is conditionally stable. This is a case where it is necessary to plot stability circles.

STABILITY CIRCLES

Let us examine the effect of Z_2 on the input impedance Z_e or, alternatively, the effect of Γ_2 on Γ_e. We know from Eqn (2.10) that

$$\Gamma_e = S_{11} + \frac{S_{12}S_{21}}{1/\Gamma_2 - S_{22}}$$

If $|\Gamma_e|$ is less than unity, there is no risk of oscillation; if $|\Gamma_e|$ is greater than unity, this is no longer the case.

Let us try to find the locus of the Γ_2 values giving the critical value $|\Gamma_e| = 1$, or

$$\left| S_{11} + \frac{S_{12}S_{21}}{1/\Gamma_2 - S_{22}} \right| = 1$$

Expanding this expression yields

$$|S_{11}(1 - \Gamma_2 S_{22}) + S_{12}S_{21}\Gamma_2| = |1 - \Gamma_2 S_{22}|$$

and introducing

$$\Delta = S_{11}S_{22} - S_{12}S_{21}$$

gives

$$|S_{11} - \Delta\Gamma_2| = |1 - \Gamma_2 S_{22}|$$

Squaring, we have

$$(S_{11} - \Delta\Gamma_2)(S_{11}^* - \Delta^*\Gamma_2^*) = (1 - \Gamma_2 S_{22})(1 - \Gamma_2^* S_{22}^*)$$

On expansion we have

$$|S_{11}|^2 + |\Delta|^2|\Gamma_2|^2 - \Delta\Gamma_2 S_{11}^* - \Delta^*\Gamma_2^* S_{11} = 1 + |\Gamma_2|^2|S_{22}|^2 - \Gamma_2 S_{22} - \Gamma_2^* S_{22}^*$$

and rearranging gives

$$(|S_{22}|^2 - |\Delta|^2)|\Gamma_2|^2 - 2\,\mathrm{Re}(\Gamma_2 S_{22}) + 2\,\mathrm{Re}(\Delta\Gamma_2 S_{11}^*) = |S_{11}|^2 - 1$$

or, again,

$$|\Gamma_2|^2 - \frac{2}{|S_{22}|^2 - |\Delta|^2}\,\mathrm{Re}\{(S_{22} - \Delta S_{11}^*)\Gamma_2\} = \frac{|S_{11}|^2 - 1}{|S_{22}|^2 - |\Delta|^2} \qquad (2.18)$$

This is the equation of a circle in the complex plane. If we refer to Fig. 2.3 we have

$$\overline{\mathrm{OM}} = \overline{\mathrm{OC}_2} + \overline{\mathrm{C}_2\mathrm{M}}$$

$$\overline{\mathrm{C}_2\mathrm{M}} = \overline{\mathrm{OM}} - \overline{\mathrm{OC}_2}$$

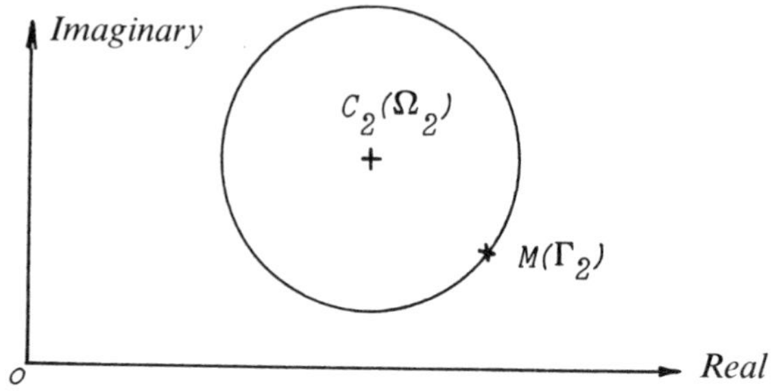

Fig. 2.3 *Circle in a complex plane.*

or in modulus terms

$$|\overline{C_2M}|^2 = |\overline{OM} - \overline{OC_2}|^2$$

$$R_2{}^2 = (\Gamma_2 - \Omega_2)(\Gamma_2{}^* - \Omega_2{}^*)$$

Hence by expansion we obtain

$$|\Gamma_2|^2 - 2\mathrm{Re}(\Omega_2{}^*\Gamma_2) = R_2{}^2 - |\Omega_2|^2 \tag{2.19}$$

where Re() indicates the real part of (). By equating with (2.18) we obtain

$$\Omega_2{}^* = \frac{S_{22} - \Delta S_{11}{}^*}{|S_{22}|^2 - |\Delta|^2}$$

$$R_2{}^2 = \frac{|S_{11}|^2 - 1}{|S_{22}|^2 - |\Delta|^2} + \frac{(S_{22} - \Delta S_{11}{}^*)(S_{22} - \Delta S_{11}{}^*)^*}{(|S_{22}|^2 - |\Delta|^2)^2}$$

Let us simplify the expression for the radius R_2. Referring to Eqn (2.13) we note that

$$(S_{22} - \Delta S_{11}{}^*)(S_{22} - \Delta S_{11}{}^*)^* = |C_2|^2$$

The expansion of this expression, apart from a permutation of subscripts, is the same as that obtained for $|C_1|^2$ in the derivation of relation (2.14):

$$|C_2|^2 = |S_{22}|^2 - |S_{11}|^2|S_{22}|^2 + |S_{12}|^2|S_{21}|^2 - |\Delta|^2 + |\Delta|^2|S_{11}|^2$$

Moreover,

$$(|S_{11}|^2 - 1)(|S_{22}|^2 - |\Delta|^2) = -|S_{22}|^2 + |S_{11}|^2|S_{22}|^2 + |\Delta|^2 - |\Delta|^2|S_{11}|^2$$

Hence

$$R_2{}^2 = \frac{|S_{12}|^2|S_{21}|^2}{(|S_{22}|^2 - |\Delta|^2)^2}$$

Rearranging the results gives the locus of Γ_2 such that $|\Gamma_e| = 1$ = circle 2:

$$\text{radius } R_2 = \frac{|S_{12}S_{21}|}{||S_{22}|^2 - |\Delta|^2|}$$

$$\text{centre } \Omega_2 = \frac{(S_{22} - \Delta S_{11}{}^*)^*}{|S_{22}|^2 - |\Delta|^2} \tag{2.20}$$

Similarly, we obtain the locus of Γ_1 such that $|\Gamma_s| = 1$ = circle 1:

$$\text{radius } R_1 = \frac{|S_{12}S_{21}|}{||S_{11}|^2 - |\Delta|^2|}$$
$$\text{centre } \Omega_1 = \frac{(S_{11} - \Delta S_{22}^*)^*}{|S_{11}|^2 - |\Delta|^2} \qquad (2.21)$$

Let us consider relations (2.20) and (2.21).

Circle 2

Consider the practical case where $|S_{11}|$ is less than unity. If $Z_2 = R_0$ then $\Gamma_2 = 0$, which corresponds to the centre of the Smith chart, and we obtain $|\Gamma_e| = |S_{11}| < 1$.

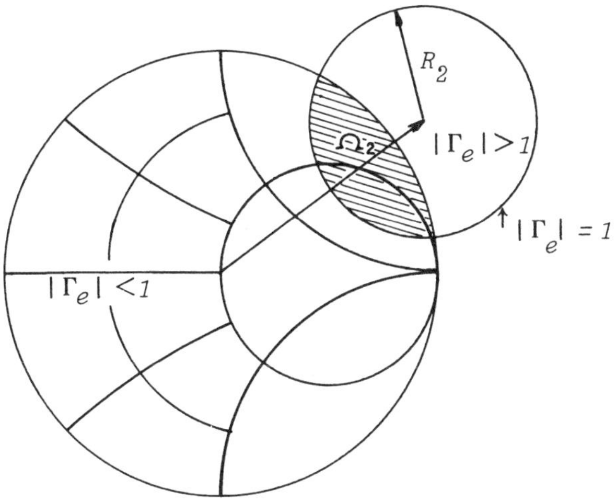

Fig. 2.4 Critical instability circle not containing the centre of the chart: $|S_{11}| < 1$.

If circle 2 does not contain the centre of the chart, the area which it delimits corresponds to values of Γ_2 causing instability; this is illustrated in Fig. 2.4 (Γ_2 must not be in the shaded area). Moreover, any risk of instability is eliminated if the inequality

$$|\Omega_2| - R_2 > 1 \qquad (2.22)$$

is satisfied.

If circle 2 contains the centre of the chart, the area which it delimits is a stable region; this is illustrated in Fig. 2.5. The same remarks apply: Γ_2 must not be in the shaded area, and any risk of instability is eliminated if

$$R_2 - |\Omega_2| > 1 \qquad (2.23)$$

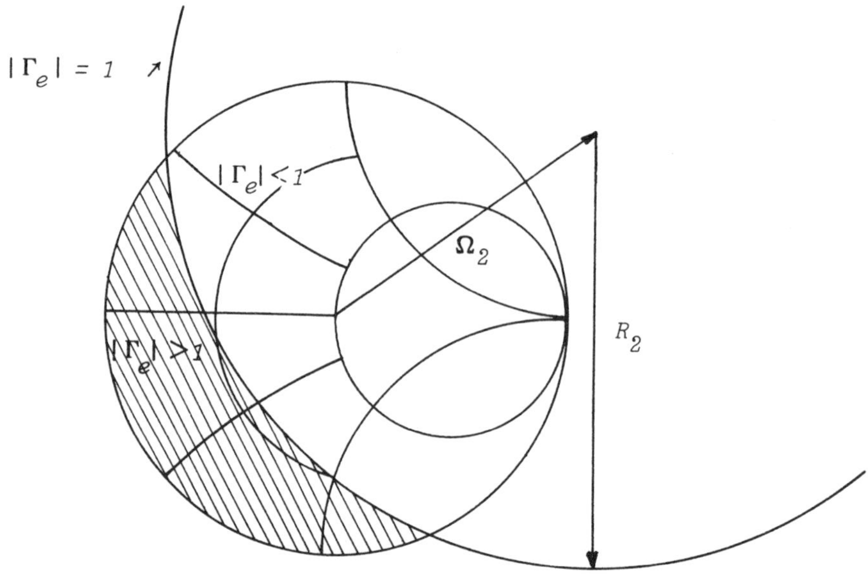

Fig. 2.5 *Critical instability circle containing the centre of the chart:* $|S_{11}|<1$.

is satisfied. Rearranging (2.22) and (2.23), we can see that there is unconditional stability at the input if

$$|R_2-|\Omega_2||>1$$

or, more explicitly,

$$\left|\frac{|S_{12}S_{21}|-|C_2{}^*|}{|S_{22}|^2-|\Delta|^2}\right|>1 \tag{2.24}$$

Circle 1

It is necessary to perform a similar analysis (case where $|S_{22}|$ is less than unity), and unconditional stability is obtained at the output if the inequality

$$\left|\frac{|S_{12}S_{21}|-|C_1{}^*|}{|S_{11}|^2-|\Delta|^2}\right|>1 \tag{2.25}$$

is satisfied.

To conclude this analysis of instability circles, let us examine the limiting case where they are tangential to the circumference of the Smith chart.

In the case of circles 2 not containing the centre of the chart, we have the inequality $|\Omega_2|-R_2>0$ which corresponds to

$$|C_2|^2>|S_{12}|^2|S_{21}|^2$$

where

$$|C_2|^2 = |S_{22}|^2 - |S_{11}|^2|S_{22}|^2 + |S_{12}|^2|S_{21}|^2 - |\Delta|^2 + |\Delta|^2|S_{11}|^2$$

(see calculation of $|C_1|^2$). Hence

$$(|S_{22}|^2 - |\Delta|^2)(1 - |S_{11}|^2) > 0$$

or, finally, if it is assumed that $|S_{11}|$ is less than unity, we have $|\Delta| < |S_{22}|$ which is the condition for $|\Omega_2| - R_2 > 0$. Thus inequality (2.22) can be expressed as

$$|C_2|^2 - (|S_{22}|^2 - |\Delta|^2 + |S_{12}S_{21}|)^2 > 0$$

or

$$|S_{22}|^2 - |\Delta|^2 - |S_{11}|^2(|S_{22}|^2 - |\Delta|^2) + |S_{12}|^2|S_{21}|^2$$
$$- (|S_{22}|^2 - |\Delta|^2)\{(|S_{22}|^2 - |\Delta|^2) + 2|S_{12}S_{21}|\} - |S_{12}|^2|S_{21}|^2 > 0$$

Hence, simplifying by means of the positive quantity $|S_{22}|^2 - |\Delta|^2$, we again find the condition

$$\frac{1 + |\Delta|^2 - |S_{11}|^2 - |S_{22}|^2}{2|S_{12}||S_{21}|} = K > 1$$

which implies $|\Omega_2| - R_2 > 1$.

In the case of circles 2 containing the centre of the chart we have $R_2 > |\Omega_2|$ and hence $|\Delta| > |S_{22}|$. Using formula (2.23), we again find $K > 1$, which will mean that $R_2 - |\Omega_2| > 1$.

The analysis for circles 1 is similar; it will now be necessary to compare $|\Delta|$ with $|S_{11}|$ by considering the practical case where $|S_{22}|$ is less than unity.

In conclusion, when $|K|$ is less than unity, it is necessary at least to examine how the instability circles (circles 1 on the one hand and circles 2 on the other) vary in the passband of the amplifier.

2.3. Unilateral transistor model

An active two-port network is termed **unilateral** when $S_{12} = 0$.

Now, referring to formula (1.45), we have

$$S_{12} = \frac{-2y_{12}}{(1 + y_{11})(1 + y_{22}) - y_{12}y_{21}}$$

and therefore $S_{12} = 0 \Leftrightarrow y_{12} = 0$. To describe a transistor as unilateral is tantamount to saying that it is **neutrodyne**.

We recall the expression for Y_{12e} (common emitter circuit) in terms of the parameters of Giacoletto's natural diagram:

$$Y_{12e} = \frac{-y_{b'c}}{1 + (y_{b'e} + y_{b'c})r_{bb'}}$$

where

$$y_{b'e} = \frac{1}{r_{b'e}} + jC_{b'e}\omega$$

$$y_{b'c} = \frac{1}{r_{b'c}} + jC_{b'c}\omega$$

We take

$$Y_{12e} \approx -\frac{jC_{b'c}\omega}{1 + r_{bb'}/r_{b'e} + jr_{bb'}(C_{b'e} + C_{b'c})\omega} \tag{2.26}$$

Thus $|S_{12e}| \to 0$ when $c_{b'c} \to 0$.

A neutralization can be characterized directly using (2.26), whence we derive

$$\frac{1}{Y_{12e}} = -\left[r_{bb'}\frac{C_{b'e} + C_{b'c}}{C_{b'c}} + \frac{1}{j\{C_{b'c}/(1 + r_{bb'}/r_{b'c})\}\omega}\right]$$

Moreover, if N is the transformation ratio of the neutralizing transformer, the compensation circuit (resistor R_n in series with capacitor C_n) will satisfy

$$R_n + \frac{1}{jC_n\omega} = \frac{-1}{NY_{12e}}$$

Hence

$$R_n = r_{bb'}\frac{C_{b'e} + C_{b'c}}{C_{b'c}}\frac{1}{N}$$

$$C_n = \frac{NC_{b'c}}{1 + r_{bb'}/r_{b'e}} \tag{2.27}$$

where N is the ratio of the primary to the secondary of the transformer.

As a final comment on transistor amplifiers used at high frequencies, let us note that manufacturers endeavour to fabricate unilateral components. Thus, for example, for an HFET 2202 with an operating bias of $\{V_{DS} = 3.5\,\text{V};\ I_{DS} = 7.5\,\text{mA}\}$, the manufacturer specifies $|S_{21}| = 4.9\,\text{dB}$ and $|S_{12}| = -24.6\,\text{dB}$, giving a non-reciprocity of modulus $|d| = -29.5\,\text{dB}$ at 4 GHz. At 8 GHz $|S_{21}| = 3.3\,\text{dB}$ and $|S_{12}| = -23.2\,\text{dB}$, and hence $|d| = -26.5\,\text{dB}$. The non-reciprocities for a bias arrangement that ensures optimum gain become $-38.35\,\text{dB}$ and $-32.6\,\text{dB}$ respectively.

Let us now consider the unilateral power gain G_{TU} of the transistor. Using expression (2.6) with $S_{12} = 0$, we obtain

$$G_{TU} = \frac{|S_{21}|^2(1 - |\Gamma_1|^2)(1 - |\Gamma_2|^2)}{|1 - \Gamma_1 S_{11}|^2 |1 - \Gamma_2 S_{22}|^2}$$

or

$$G_{TU} = \frac{1-|\Gamma_1|^2}{|1-\Gamma_1 S_{11}|^2}|S_{21}|^2\frac{1-|\Gamma_2|^2}{|1-\Gamma_2 S_{22}|^2} \tag{2.28}$$

G_{TU} is therefore given by the product of three terms:

$$G_1 = \frac{1-|\Gamma_1|^2}{|1-\Gamma_1 S_{11}|^2} \tag{2.29}$$

which represents the 'gain' due to the input matching circuit,

$$G_0 = |S_{21}|^2$$

which is the forward power gain of the transistor alone, i.e. assumed to be inserted between two 50 Ω resistances, and

$$G_2 = \frac{1-|\Gamma_2|^2}{|1-\Gamma_2 S_{22}|^2} \tag{2.30}$$

which represents the gain due to the output matching circuit.

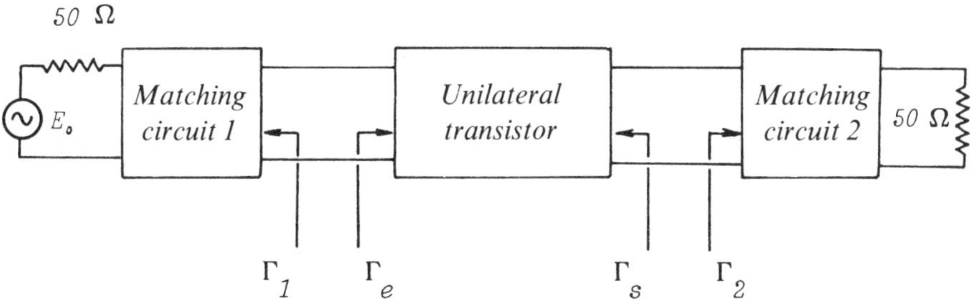

Fig. 2.6 *Matching of a unilateral transistor.*

If we assume that S_{11} and S_{22} have a modulus of less than unity, it is obvious that the unilateral transistor is unconditionally stable. In fact, when $|S_{12}| = 0$, circles 2 and 1 defined by (2.20) and (2.21) reduce to a point $\Omega_i = 1/S_{ii}$ $(i = 1, 2)$. It is therefore always possible to achieve simultaneous matching at input and output. A two-port network of this kind inserted between two matching circuits with reflection coefficients Γ_1 and Γ_2 can be represented as shown in Fig. 2.6. In this case

$$\Gamma_e = S_{11} + \frac{S_{12}S_{21}}{1/\Gamma_2 - S_{22}} = S_{11}$$

$$\Gamma_s = S_{22} + \frac{S_{12}S_{21}}{1/\Gamma_1 - S_{11}} = S_{22}$$

In order to achieve simultaneous matching, it is therefore merely necessary to satisfy

$$\begin{aligned} \Gamma_1 &= S_{11}^* \\ \Gamma_2 &= S_{22}^* \end{aligned} \tag{2.31}$$

for $|S_{11}| < 1$ and $|S_{22}| < 1$.

The maximum transducer power gain is therefore

$$G_{TU(max)} = \frac{1}{1-|S_{11}|^2}|S_{21}|^2\frac{1}{1-|S_{22}|^2} \tag{2.32a}$$

or

$$G_{TU(max)} = G_{1(max)}G_0G_{2(max)}$$

where, of course,

$$G_{1(max)} = \frac{1}{1-|S_{11}|^2} \quad G_{2(max)} = \frac{1}{1-|S_{22}|^2} \tag{2.32b}$$

Note that, without a matching circuit, we have $\Gamma_1 = \Gamma_2 = 0$ and therefore $G_1 = G_2 = 1$, or

$$G_{TU} = G_0$$

The quantity $(G_{1(max)})_{dB} + (G_{2(max)})_{dB}$ represents the additional gain which we may expect from the circuit when perfect matching is achieved. Conversely, we may also have $|\Gamma_1| = |\Gamma_2| = 1$ and hence $G_1 = G_2 = 0$; this means that G_1 and G_2 both vary between zero and a maximum value. For example, for G_1 we have

$$0 \leqslant G_1 \leqslant G_{1(max)} = \frac{1}{1-|S_{11}|^2}$$

Let us now consider the following problem: what is the locus of the reflection coefficients Γ_1 at the input such that the gain G_1 is constant?

We have to solve the equation

$$\frac{1-|\Gamma_1|^2}{|1-\Gamma_1 S_{11}|^2} = G_1 = \text{constant} \tag{2.33}$$

which, on expansion, becomes

$$1-|\Gamma_1|^2 = G_1(1-\Gamma_1 S_{11})(1-\Gamma_1^* S_{11}^*)$$

or

$$1-|\Gamma_1|^2 = G_1 + G_1|\Gamma_1|^2|S_{11}|^2 - 2G_1 \operatorname{Re}(\Gamma_1 S_{11})$$

Rearranging this expression gives

$$|\Gamma_1|^2(1+G_1|S_{11}|^2) - 2G_1 \operatorname{Re}(\Gamma_1 S_{11}) = 1 - G_1$$

We therefore find (cf. Eqn (2.19)) the equation of a circle:

$$|\Gamma_1|^2 - 2\frac{G_1}{1+G_1|S_{11}|^2}\mathrm{Re}(S_{11}\Gamma_1) = \frac{1-G_1}{1+G_1|S_{11}|^2} \tag{2.34}$$

with centre C_{G_1} (Ω_{G_1}) and radius R_{G_1} such that

$$\Omega_{G_1}{}^* = \frac{G_1 S_{11}}{1+G_1|S_{11}|^2}$$

$$R_{G_1}{}^2 = \frac{(1-G_1)(1+G_1|S_{11}|^2)+G_1{}^2|S_{11}|^2}{(1+G_1|S_{11}|^2)^2}$$

Hence the locus of Γ_1 such that G_1 = constant is circle G_1 with

$$\text{radius } R_{G_1} = \frac{\{1-G_1(1-|S_{11}|^2)\}^{\frac{1}{2}}}{1+G_1|S_{11}|^2}$$

$$\text{centre } \Omega_{G_1} = \frac{G_1}{1+G_1|S_{11}|^2}S_{11}{}^* \tag{2.35}$$

The centre of circle G_1 therefore passes along the straight line segment connecting the centre of the Smith chart (for $G_1 = 0$ and therefore $R_{G_1} = 1$) to the point representing $S_{11}{}^*$ (for $G_1 = G_{1(\max)}$ and therefore $R_{G_1} = 0$).

On Fig. 2.7 the constant-gain circles are plotted for five values of G_1 between 1 and $G_{1(\max)}$. For $G_1 = 1$, it can be shown that

$$R_{G_1} = |\Omega_{G_1}| = \frac{|S_{11}|}{1+|S_{11}|^2}$$

Therefore circle $G_1 = 0\,\text{dB}$ indeed passes through the centre of the chart.

Note

We can express relations (2.35) differently by introducing the normalized gain

$$g_1 = \frac{G_1}{G_{1(\max)}} = G_1(1-|S_{11}|^2) \tag{2.36}$$

Then

$$1-g_1 = 1-G_1+G_1|S_{11}|^2$$

Hence

$$1+G_1|S_{11}|^2 = 1-g_1+\frac{g_1}{1-|S_{11}|^2} = \frac{1-|S_{11}|^2(1-g_1)}{1-|S_{11}|^2}$$

and therefore the locus of Γ_1 such that G_1 = constant is the circle g_1 with

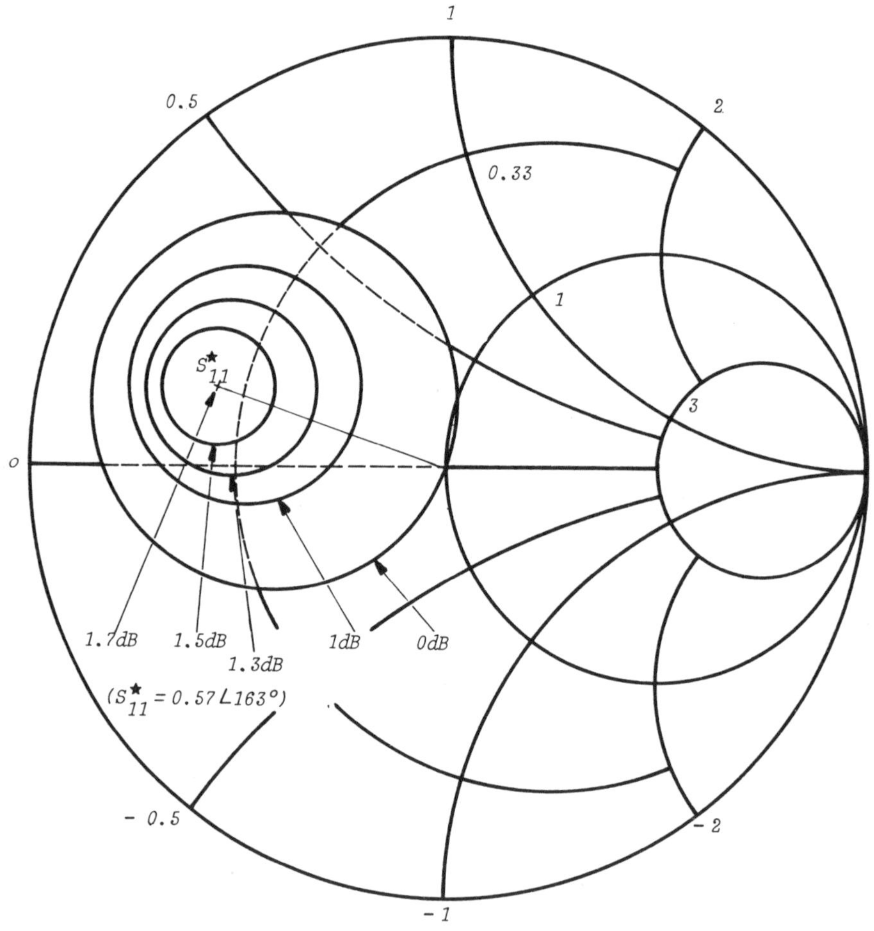

Fig. 2.7 *Constant-gain circles at the input of a unilateral transistor.*

$$\text{radius } R_{g_1} = \frac{(1-g_1)^{\frac{1}{2}}(1-|S_{11}|^2)}{1-|S_{11}|^2(1-g_1)}$$

$$\text{centre } \Omega_{g_1} = \frac{g_1 S_{11}^*}{1-|S_{11}|^2(1-g_1)} \tag{2.37}$$

At the output side we obtain the same set of equations; for example the locus of Γ_2 such that g_2 = constant is the circle g_2 with

$$\text{radius } R_{g_2} = \frac{(1-g_2)^{\frac{1}{2}}(1-|S_{22}|^2)}{1-|S_{22}|^2(1-g_2)}$$

$$\text{centre } \Omega_{g_2} = \frac{g_2 S_{22}^*}{1-|S_{22}|^2(1-g_2)} \tag{2.38}$$

where

$$g_2 = \frac{G_2}{G_{2(\max)}} = G_2(1-|S_{22}|^2)$$

It is unquestionably at the input that constant-gain circles assume great importance, as they make it possible to find a compromise with as low a noise factor as possible (we shall consider this topic in detail later).

In conclusion, we may wish to know the error in the calculation of the transducer power gain when S_{12} is disregarded, i.e. we may wish to compare the G_T of Eqn (2.6) with the G_{TU} of Eqn (2.28). We therefore introduce the figure of merit

$$u = \frac{|\Gamma_1 \Gamma_2 S_{12} S_{21}|}{|(1-\Gamma_1 S_{11})(1-\Gamma_2 S_{22})|}$$

and use the double inequality

$$\frac{1}{(1+u)^2} < \frac{G_T}{G_{TU}} < \frac{1}{(1-u)^2} \tag{2.39}$$

2.4. Constant-noise-factor circles

We recall that the noise factor represents the degradation of the signal-to-noise ratio between the input and output of the two-port network:

$$F = \frac{(S/N)_{\text{in}}}{(S/N)_{\text{out}}}$$

If G is the power gain, we can write

$$F = \frac{S_e/N_{in}}{GS_e/G(N_{in}+N_a)}$$

where N_a is the additional noise power of the amplifier referred to the input. We finally obtain

$$F = 1 + \frac{N_a}{N_{in}} \tag{2.40}$$

We propose to calculate the quantity N_a/N_{in} by representing a noisy two-port network by a noiseless two-port network with a noise voltage source v connected in series at its input and a current generator i connected in parallel, as shown in Fig. 2.8 where $Z_s = R_s + jX_s$ is the source impedance with which we associate the noise voltage v_s and Z_L is the load impedance.

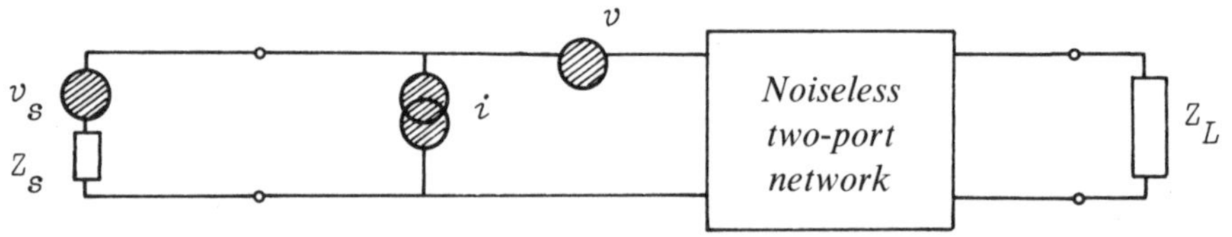

Fig. 2.8 Model of a noisy two-port network.

We also recall the thermal noise equation, or **Nyquist formula**, which expresses the mean quadratic value of the noise voltage v_{th} across a resistance R:

$$\overline{v_{th}^2} = 4kTR\Delta f$$

where k is the Boltzmann constant (1.38054×10^{-23} J K^{-1}), T is the absolute temperature and Δf is the bandwidth under consideration.

We therefore have

$$\overline{v_s^2} = 4kTR_s\Delta f \tag{2.41}$$

where R_s is the real part of Z_s. Similarly, we can write

$$\overline{v^2} = 4kTR_n\Delta f$$

$$\overline{i^2} = 4kTG_p\Delta f \tag{2.42}$$

where R_n and G_p are the **equivalent noise resistance** and **equivalent noise conductance** respectively of the two-port network.

Let us now introduce, as an intermediary in the calculation, the input impedance $Z = R + jX$ of the two-port network. The noise voltages v_s' and v'

across this impedance, which are due to v_s and v respectively, are given by

$$v_s' = v_s \frac{Z}{R+R_s+j(X+X_s)}$$

$$v' = v \frac{Z}{R+R_s+j(X+X_s)}$$

Current i causes a current i_R to flow in R such that

$$(R+jX)i_R = (i-i_R)(R_s+jX_s)$$

or

$$i_R = i \frac{Z_s}{R+R_s+j(X+X_s)}$$

Therefore the noise voltage v_i' across Z due to current i is given by

$$v_i' = i \frac{ZZ_s}{R+R_s+j(X+X_s)}$$

Note that the mean quadratic values of the three voltages calculated in this way are proportional to the quantity

$$\frac{|Z|^2}{(R+R_s)^2+(X+X_s)^2}$$

We are now in a position to obtain the noise powers N_{in} and N_a which appear in Eqn (2.40):

$$N_{in} = \overline{v_s'^2}\,\mathrm{Re}\left(\frac{1}{Z}\right)$$

$$N_a = \overline{(v'+v_i')^2}\,\mathrm{Re}\left(\frac{1}{Z}\right)$$

Note that, under matching conditions, we have $\mathrm{Re}(1/z) = \mathrm{Re}(1/z_s)$. Thus

$$\frac{N_a}{N_{in}} = \frac{\overline{(v'+v_i')^2}}{\overline{v_s'^2}}$$

After simplification using the proportionality coefficient mentioned above, we obtain the following expression for the noise factor:

$$F = 1 + \frac{\overline{(v+Z_s i)(v+Z_s i)^*}}{\overline{v_s^2}} \tag{2.43}$$

The numerator of the fraction can be expanded as follows:

$$\begin{aligned}\overline{(v+Z_s i)(v+Z_s i)^*} &= \overline{v^2} + |Z_s|^2\overline{i^2} + \overline{Z_s iv^* + Z_s^* i^* v} \\ &= \overline{v^2} + |Z_s|^2\overline{i^2} + R_s\overline{(iv^*+i^*v)} + X_s j\overline{(iv^*-i^*v)}\end{aligned}$$

We assume that the noise sources v and i are correlated, or, to be more precise, that there is a **correlation impedance** $Z_c = R_c + jX_c$ such that

$$v = v_n + Z_c i$$

Then

$$\overline{iv^* + i^*v} = 2R_c\overline{i^2}$$

$$\overline{j(iv^* - i^*v)} = 2X_c\overline{i^2}$$

Hence the expanded form of Eqn (2.43) is

$$F = 1 + \frac{\overline{v^2} + |Z_s|^2\overline{i^2} + 2R_sR_c\overline{i^2} + 2X_sX_c\overline{i^2}}{\overline{v_s^{\,2}}}$$

By using Eqns (2.41) and (2.42) this expression can also be written in the form

$$F = 1 + \frac{R_n}{R_s} + \frac{G_p}{R_s}(R_s^{\,2} + X_s^{\,2}) + 2R_cG_p + 2X_cG_p\frac{X_s}{R_s} \qquad (2.44)$$

The noise factor of a two-port network is therefore a function of the source impedance Z_s. It is easy to demonstrate the existence of an optimum impedance $Z_s = Z_{opt}$ for which the noise factor is minimum and equal to F_{min}; by equating the partial derivatives to zero

$$\frac{\partial F}{\partial X_s} = 0$$

$$\frac{\partial F}{\partial R_s} = 0$$

we finally obtain

$$R_{opt} = \left(\frac{R_n}{G_p} - X_c^{\,2}\right)^{\frac{1}{2}}$$

$$|X_c| < \left(\frac{R_n}{G_p}\right)^{\frac{1}{2}} \qquad (2.45)$$

$$X_{opt} = -X_c$$

and of course

$$Z_{opt} = R_{opt} + jX_{opt}$$

For $R_s = R_{opt}$ and $X_s = X_{opt}$ we therefore have

$$F = F_{min} = 1 + 2R_cG_p + \frac{R_n - G_pX_c^{\,2}}{(R_n/G_p - X_c^{\,2})^{\frac{1}{2}}} + G_p\left(\frac{R_n}{G_p} - X_c^{\,2}\right)^{\frac{1}{2}}$$

and hence

$$F_{min} = 1 + 2G_p\left\{R_c + \left(\frac{R_n}{G_p} - X_c^{\,2}\right)^{\frac{1}{2}}\right\} \qquad (2.46)$$

In the expression for the noise factor given by Eqn (2.44) the unknowns are R_n, G_p, R_c and X_c. We now attempt to find a new formulation in terms of R_n, F_{min}, R_{opt} and X_{opt}. Now, from (2.45) and (2.46) we derive

$$X_c = -X_{opt}$$

$$G_p = \frac{R_n}{R_{opt}^2 + X_{opt}^2} \tag{2.47}$$

$$R_c = \frac{F_{min} - 1}{2R_n}(R_{opt}^2 + X_{opt}^2) - R_{opt}$$

We also observe that

$$\frac{R_n}{G_p} = R_{opt}^2 + X_{opt}^2 > X_c^2$$

However,

$$2R_c G_p = F_{min} - 1 - \frac{2R_n R_{opt}}{R_{opt}^2 + X_{opt}^2}$$

Thus, considering Eqn (2.44) once more, we infer that

$$F - F_{min} = \frac{R_n}{R_s(R_{opt}^2 + X_{opt}^2)}\{(R_{opt}^2 + X_{opt}^2) + (R_s^2 + X_s^2) - 2R_{opt}R_s - 2X_{opt}X_s\}$$

or

$$F = F_{min} + \frac{R_n}{R_s(R_{opt}^2 + X_{opt}^2)}\{(R_s - R_{opt})^2 + (X_s - X_{opt})^2\} \tag{2.48}$$

More concisely,

$$F = F_{min} + \frac{R_n}{R_s}\left|\frac{Z_s}{Z_{opt}} - 1\right|^2 \tag{2.49}$$

If we now introduce the admittances Y_s and Y_{opt} such that

$$Z_s = \frac{1}{Y_s} = \frac{1}{G_s + jB_s}$$

$$Z_{opt} = \frac{1}{Y_{opt}} = \frac{1}{G_{opt} + jB_{opt}}$$

Eqn (2.48) can be written in the form

$$F = F_{min} + \frac{R_n}{G_s}\{(G_s - G_{opt})^2 + (B_s - B_{opt})^2\} \tag{2.50}$$

or again

$$F = F_{min} + \frac{R_n}{G_s}\left|Y_s - Y_{opt}\right|^2 \tag{2.51}$$

Finally, if we use the normalized admittances

$$y_s = R_0 Y_s = g_s + jb_s$$

and

$$y_{opt} = R_0 Y_{opt} = g_{opt} + jb_{opt}$$

we have

$$F = F_{min} + \frac{r_n}{g_s}\{(g_s - g_{opt})^2 + (b_s - b_{opt})^2\} \tag{2.52}$$

where $r_n = R_n/R_0$.

The noise factor can also be expressed in terms of the reflection coefficients Γ_s and Γ_{opt}:

$$\Gamma_s = \frac{1 - y_s}{1 + y_s}$$

$$\Gamma_{opt} = \frac{1 - y_{opt}}{1 + y_{opt}}$$

This enables us to determine, on the Smith chart, the locus of the reflection coefficients Γ_s such that the noise factor F is constant.

Starting with Eqn (2.52) we can write

$$F = F_{min} + r_n \frac{|y_s - y_{opt}|^2}{g_s} \tag{2.53}$$

where

$$y_s = \frac{1 - \Gamma_s}{1 + \Gamma_s}$$

$$y_{opt} = \frac{1 - \Gamma_{opt}}{1 + \Gamma_{opt}}$$

Hence

$$|y_s - y_{opt}|^2 = \frac{|(1 - \Gamma_s)(1 + \Gamma_{opt}) - (1 - \Gamma_{opt})(1 + \Gamma_s)|^2}{|(1 + \Gamma_s)(1 + \Gamma_{opt})|^2}$$

Expanding and simplifying, we obtain

$$|y_s - y_{opt}|^2 = 4\frac{|\Gamma_s - \Gamma_{opt}|^2}{|1 + \Gamma_s|^2 |1 + \Gamma_{opt}|^2}$$

However,

$$g_s = \frac{\mathrm{Re}\{(1 - \Gamma_s)(1 + \Gamma_s^*)\}}{|1 + \Gamma_s|^2} = \frac{1 - |\Gamma_s|^2}{|1 + \Gamma_s|^2}$$

and thus

$$\frac{|y_s - y_{opt}|^2}{g_s} = 4\frac{|\Gamma_s - \Gamma_{opt}|^2}{(1-|\Gamma_s|^2)|1+\Gamma_{opt}|^2}$$

We finally obtain the expression for F in terms of the reflection coefficients Γ_s and Γ_{opt}:

$$F = F_{min} + 4r_n\frac{|\Gamma_s - \Gamma_{opt}|^2}{(1-|\Gamma_s|^2)|1+\Gamma_{opt}|^2} \tag{2.54}$$

Let us stress the fact that $r_n = R_n/R_0$.

It should be noted that we can determine F_{min} experimentally by the method of successive approximations and that therefore $\Gamma_s = \Gamma_{opt}$. Given a specific value of Γ_s we can then determine r_n; for example, for $\Gamma_s = 0$ (i.e. $Z_s = R_0$) we obtain

$$F_0 = F_{min} + 4r_n\frac{|\Gamma_{opt}|^2}{|1+\Gamma_{opt}|^2}$$

and hence

$$r_n = (F_0 - F_{min})\frac{|1+\Gamma_{opt}|^2}{4|\Gamma_{opt}|^2} \tag{2.55}$$

We recall that our intention was to find the position on the Smith chart of the reflection coefficients Γ_s such that F = constant. For this purpose we introduce the quantity

$$N = \frac{F - F_{min}}{4r_n}|1+\Gamma_{opt}|^2$$

In terms of N, Eqn (2.54) can be rewritten in the form

$$N(1-|\Gamma_s|^2) = |\Gamma_s - \Gamma_{opt}|^2$$

which, on expansion, becomes

$$N - N|\Gamma_s|^2 = |\Gamma_s|^2 + |\Gamma_{opt}|^2 - 2\text{Re}(\Gamma_{opt}{}^*\Gamma_s)$$

On rearranging the terms, we find an expression for a circle (cf. Eqn (2.19)):

$$|\Gamma_s|^2 - \frac{2}{1+N}\text{Re}(\Gamma_{opt}{}^*\Gamma_s) = \frac{N-|\Gamma_{opt}|^2}{1+N}$$

This therefore yields the following result: the locus of Γ_s such that F is a constant is a circle N with

$$\text{centre } \Omega_N = \frac{\Gamma_{opt}}{1+N}$$

$$\text{radius } R_N = \frac{\{N^2 + N(1-|\Gamma_{opt}|^2)\}^{\frac{1}{2}}}{1+N} \tag{2.56}$$

where

$$N = \frac{F - F_{min}}{4r_n}|1 + \Gamma_{opt}|^2$$

Note that $N = 0$ for $F = F_{min}$ and the locus reduces to the point Γ_{opt}.

Figure 2.9 illustrates the relations given above. We have drawn the noise circles of a microwave transistor with an operating bias $\{V_{CE} = 10\,V; I_c = 4\,mA\}$ for which the manufacturer specifies

$$F_{min} = 1.8\,dB \quad (1.5136)$$

$$R_n = 13.8\,\Omega \quad (r_n = 0.276)$$

$$\Gamma_{opt} = 0.412 \quad (\text{angle } 104°)$$

at 2.5 GHz.

Our examination of the noise factor of a two-port network would not be complete without mention of **Friis' formula** from which we can derive two quantities, the **minimum noise figure** and the **figure of merit**. These quantities are very useful in practice.

Let us recall that Friis' formula enables us to calculate the noise factor F_T of a cascaded n-stage amplifier:

$$F_T = F_1 + \frac{F_2 - 1}{G_1} + \frac{F_3 - 1}{G_1 G_2} + \dots + \frac{F_n - 1}{G_1 G_2 \dots G_{n-1}} \tag{2.57}$$

where F_i is the noise factor of the ith stage and G_i is the available power gain of this stage. Let us assume that these n stages are identical and are designed to provide the minimum noise factor; then

$$F_1 = \dots = F_i = \dots = F_n = F_{min}$$

and the associated gain is

$$G_1 = \dots = G_i = \dots = G_n = G_a$$

Equation (2.57) can then be written as

$$F_{T(min)} = 1 + (F_{min} - 1)\left(1 + \frac{1}{G_a} + \frac{1}{G_a^{\,2}} + \dots + \frac{1}{G_a^{\,n-1}}\right)$$

and if the number of stages tends to infinity

$$\underset{n \to \infty}{F_{T(min)}} \to M_{min} = 1 + \frac{F_{min} - 1}{1 - 1/G_a} \tag{2.58}$$

This is termed the minimum noise figure. M_{min} is frequently expressed in decibels.

Example

Let there be a transistor for which $F_{min} = 1.4\,dB$ and $G_a = 14\,dB$ at 1.5 GHz; we therefore obtain $M_{min} = 1.45\,dB$.

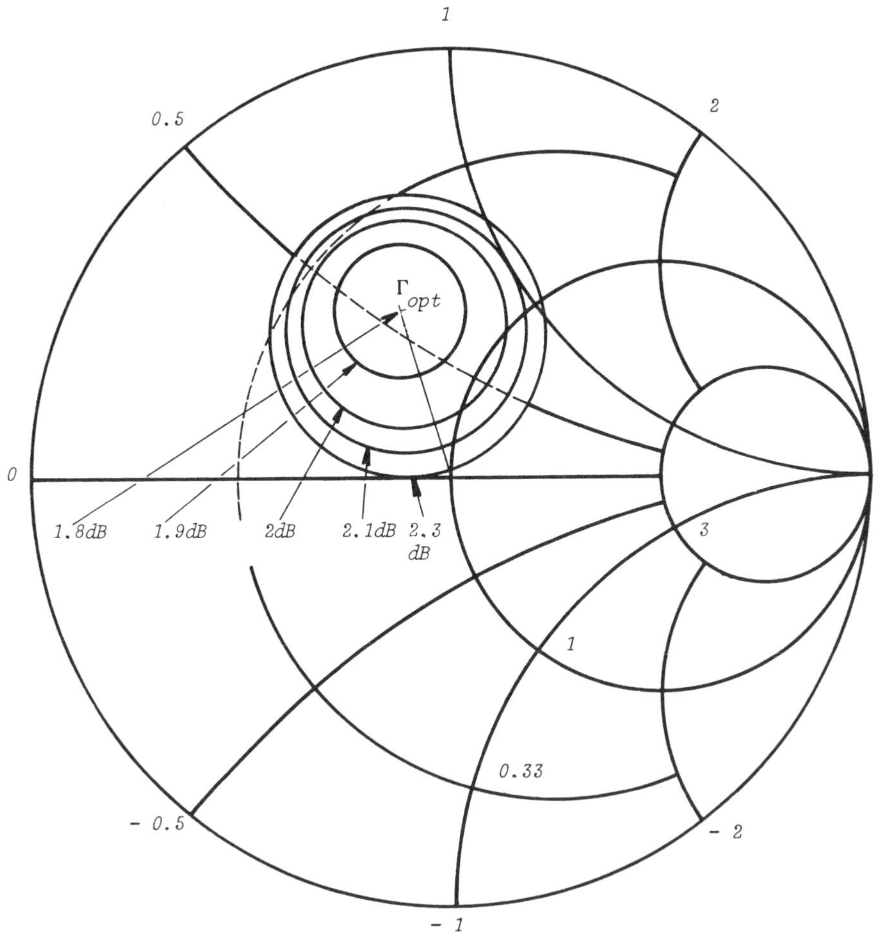

Fig. 2.9 *Noise circles of a transistor used at 2.5 GHz.*

Let us now apply Friis' formula to the case of two amplifiers A and B. If A is the leading amplifier, we have

$$F_{AB} = F_A + \frac{F_B - 1}{G_A}$$

If B is the leading amplifier, we then have

$$F_{BA} = F_B + \frac{F_A - 1}{G_B}$$

Let us assume that configuration AB provides the lowest noise factor; therefore

$$F_{AB} < F_{BA}$$

or

$$F_A + \frac{F_B - 1}{G_A} < F_B + \frac{F_A - 1}{G_B}$$

Hence

$$(F_A - 1) - \frac{F_A - 1}{G_B} < (F_B - 1) - \frac{F_B - 1}{G_A}$$

and finally

$$\frac{F_A - 1}{1 - 1/G_A} < \frac{F_B - 1}{1 - 1/G_B}$$

which we can write in the form

$$M_A < M_B$$

where

$$M = \frac{F - 1}{1 - 1/G} \tag{2.59}$$

M is the figure of merit of an amplifier of noise factor F and gain G; it is not desirable to express this in decibels.

Note

The figure of merit is only a function of the reflection coefficient Γ_s of the source. From (2.54) we obtain

$$F - 1 = \frac{1}{|1 + \Gamma_{opt}|^2} \frac{(F_{min} - 1)|1 + \Gamma_{opt}|^2(1 - |\Gamma_s|^2) + 4r_n|\Gamma_s - \Gamma_{opt}|^2}{1 - |\Gamma_s|^2}$$

Using the expression for the transducer power gain G_T derived from (2.6), i.e.

$$G_T = |S_{21}|^2 \frac{1 - |\Gamma_s|^2}{|1 - \Gamma_s S_{11}|^2} \frac{1 - |\Gamma_L|^2}{|1 - \Gamma_L S'_{22}|^2}$$

in which

$$S'_{22} = \frac{S_{22} - \Delta\Gamma_s}{1 - S_{11}\Gamma_s}$$

we infer that, when the output is matched ($\Gamma_L = S'_{22}{}^*$),

$$G_A = |S_{21}|^2 \frac{1 - |\Gamma_s|^2}{|1 - S_{11}\Gamma_s|^2 - |S_{22} - \Delta\Gamma_s|^2}$$

and hence

$$1 - \frac{1}{G_A} = \frac{1}{|S_{21}|^2} \frac{(1 - |\Gamma_s|^2)|S_{21}|^2 + |S_{22} - \Delta\Gamma_s|^2 - |1 - S_{11}\Gamma_s|^2}{1 - |\Gamma_s|^2}$$

Consequently, Eqn (2.59) can be rewritten in the form

$$M = \frac{|S_{21}|^2}{|1 + \Gamma_{opt}|^2} \frac{(F_{min} - 1)|1 + \Gamma_{opt}|^2(1 - |\Gamma_s|^2) + 4r_n|\Gamma_s - \Gamma_{opt}|^2}{(1 - |\Gamma_s|^2)|S_{21}|^2 + |S_{22} - \Delta\Gamma_s|^2 - |1 - S_{11}\Gamma_s|^2}$$

We put

$$m = M\frac{|1 + \Gamma_{opt}|^2}{|S_{21}|^2}$$

and make use of the identity

$$|a + b|^2 \equiv |a|^2 + |b|^2 + 2\mathrm{Re}(ab^*)$$

We obtain then the very important relation

$$\{4r_n - f - m(|\Delta|^2 - |S_{11}|^2 - |S_{21}|^2)\}|\Gamma_s|^2 - 2\mathrm{Re}\{(4r_n\Gamma_{opt}{}^* + mC_1)\Gamma_s\}$$
$$= m(|S_{21}|^2 + |S_{22}|^2 - 1) - f - 4r_n|\Gamma_{opt}|^2$$

where

$$f = (F_{min} - 1)|1 + \Gamma_{opt}|^2$$

$$m = M\frac{|1 + \Gamma_{opt}|^2}{|S_{21}|^2} \qquad (2.60)$$

and, from Eqn (2.12),

$$C_1 = S_{11} - \Delta S_{22}{}^*$$

In fact, referring to (2.19), we have thus demonstrated the following original proposition: the locus of the reflection coefficients of the source ensuring a constant figure of merit is a circle.

Let us add, as a corollary, that we obtain the minimum figure of merit by writing the radius as zero. In this hypothesis, the reflection coefficient which must be presented by the source will of course be equal to the value of the point representing the centre when m is minimum.

Chapter 3

Amplifier and oscillator design

3.1. Narrow band amplifier using lumped elements

We propose to characterize an amplifier centred on 500 MHz using the BFT 65 transistor shown in Fig. 3.1. At the operating point $\{V_{CE} = 8\ \text{V};\ I_C = 25\,\text{mA}\}$ and at 500 MHz, the S parameters in the common emitter configuration have the values $S_{11e} = 0.343\,(-174\,°)$, $S_{21e} = 6.311\,(85°)$, $S_{12e} = 0.058\,(72°)$ and $S_{22e} = 0.441\,(-23°)$, yielding $|\Delta_e| = 0.216$. Then, from Eqn (2.14a),

$$K = 1.0034$$

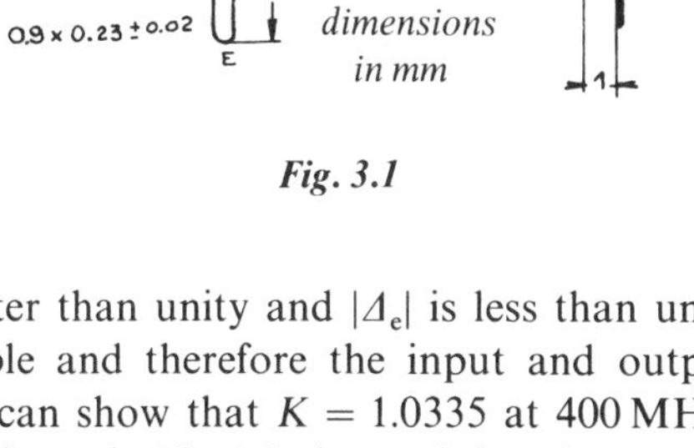

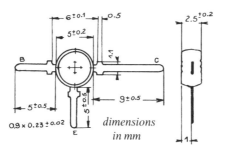

Fig. 3.1

Since K is greater than unity and $|\Delta_e|$ is less than unity, the transistor is unconditionally stable and therefore the input and output can be matched simultaneously. We can show that $K = 1.0335$ at 400 MHz and $K = 1.154$ at 600 MHz. In fact, as shown in Fig. 3.2, the modulus of S_{21e} decreases as a function of frequency. It is therefore necessary to design the amplifier at a frequency greater than 500 MHz. We shall adopt the value of 550 MHz for which $S_{11e} = 0.345\,(-177°)$, $S_{21e} = 5.774\,(82°)$, $S_{12e} = 0.063\,(72°)$ and $S_{22e} = 0.390\,(-21°)$. We then compute $K = 1.0754$ with $|\Delta_e| = 0.2312$. Thus, using (2.16), we can expect a maximum transducer power gain of

$$G_{T(\max)} = 17.95\,\text{dB}$$

In order to obtain some idea of the gains provided by the input and output

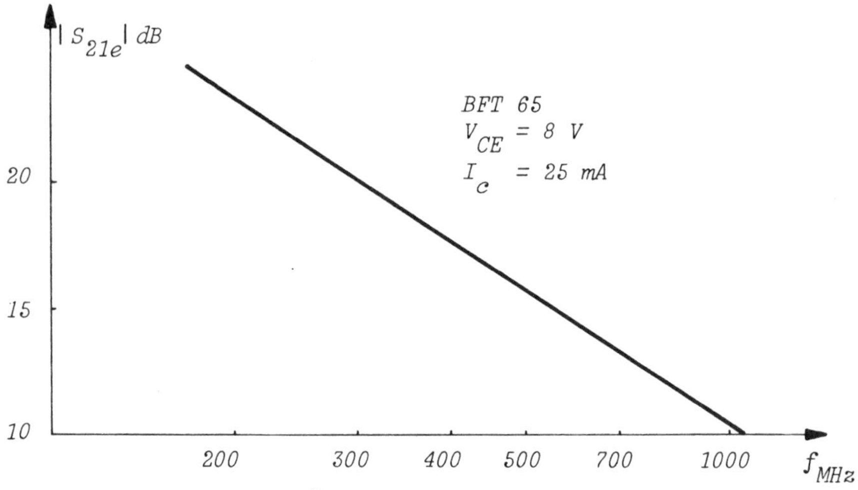

Fig. 3.2 S_{21e} *decreasing at about 6 dB per octave.*

matching circuits, we can regard the transistor as unilateral and use Eqn (2.32):

$$G_{TU(max)} = G_{1(max)}G_0G_{2(max)}$$

with

$$G_{1(max)} = 10\log\left(\frac{1}{1-|S_{11}|^2}\right) = 0.55\,\text{dB}$$

$$G_0 = 20\log|S_{21}| = 15.23\,\text{dB}$$

$$G_{2(max)} = 10\log\left(\frac{1}{1-|S_{22}|^2}\right) = 0.72\,\text{dB}$$

Hence

$$G_{TU(max)} = 16.5\,\text{dB}$$

The error resulting from regarding the transistor as unilateral is therefore 1.5 dB, which is not negligible.

The calculations for the matching circuits will be performed using Eqns (2.12) and (2.13) in which the minus sign must be used because $|\Delta|$ is less than unity. We compute

$$(B_1^2 - 4|C_1|^2)^{\frac{1}{2}} = (B_2^2 - 4|C_2|^2)^{\frac{1}{2}} = 0.2281$$

Hence

$$\Gamma_1 = 0.7213\,(180°)$$

$$\Gamma_2 = 0.7386\,(23°)$$

Implementation of the input circuit

Note that the impedance $Z_1 = R_1 + jX_1$ corresponding to Γ_1 is $Z_1 = (8.1$

$+0j)\ \Omega$. Since R_1 is less than 50 Ω, we shall use the circuit shown in Fig. 3.3(a). If we put $X = L_1\omega$, $B = C_1\omega$, taking into account that in practice $R_0 = 50\ \Omega$, the input impedance is simply

$$Z_1 = \frac{1}{1/50 + jB} + jX$$

or

$$R_1 + jX_1 = \frac{50}{1+50^2B^2} + j\left(X - \frac{50^2 B}{1+50^2B^2}\right)$$

Equating real and imaginary parts gives

$$B = C_1\omega = \frac{(50-R_1)^{\frac{1}{2}}}{50R_1^{\frac{1}{2}}} \qquad (3.1)$$

$$X = L_1\omega = X_1 + \{R_1(50-R_1)\}^{\frac{1}{2}} \qquad (3.2)$$

Numerical substitution at 550 MHz gives

$C_1\omega = 45.5\ \text{mS}$ and hence $C_1 = 13.2\ \text{pF}$

$L_1\omega = 18.4\ \Omega$ and hence $L_1 = 5.3\ \text{nH}$

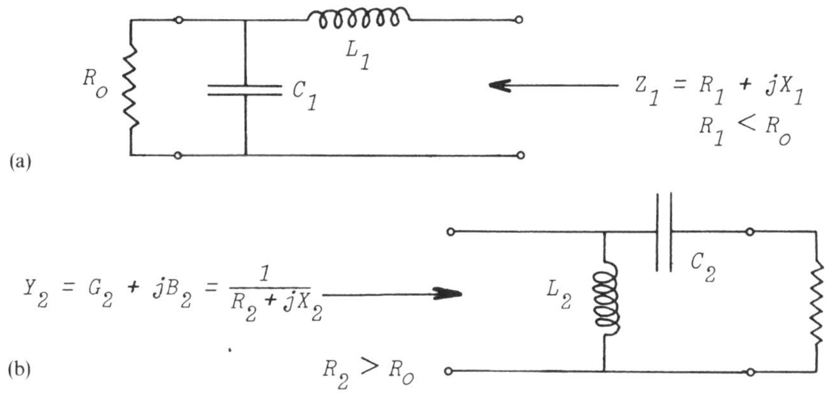

Fig. 3.3 (a) Input matching circuit; (b) output matching circuit.

Implementation of the output circuit

Impedance Z_2 corresponding to Γ_2 has a value of

$$Z_2 = R_2 + jX_2 = (122.3 + 155.4j)\ \Omega$$

As R_2 is now greater than R_0, we shall use the circuit shown in Fig. 3.3(b). The numerical value of Y_2 is $(3.127 - 3.974j)$ mS, and we write

$$Y_2 = -jB + \frac{1}{50 - jX}$$

where $B = 1/L_2\omega$ and $X = 1/C_2\omega$, or

$$G_2 + jB_2 = \frac{50}{50^2 + X^2} + j\left(\frac{X}{50^2 + X^2} - B\right)$$

which gives

$$X = \frac{1}{C_2\omega} = \left\{50\left(\frac{1}{G_2} - 50\right)\right\}^{\frac{1}{2}} \tag{3.3}$$

$$B = \frac{1}{L_2\omega} = \left\{G_2\left(\frac{1}{50} - G_2\right)\right\}^{\frac{1}{2}} - B_2 \tag{3.4}$$

Numerical substitution, again at 550 MHz, gives

$$C_2\omega = 8.6\,\text{mS} \qquad \text{and hence} \qquad C_2 = 2.5\,\text{pF}$$

$$L_2\omega = 89\,\Omega \qquad \text{and hence} \qquad L_2 = 26\,\text{nH}$$

Note

It is also possible to use a Smiths chart to design these matching circuits. The principle consists in combining the impedance chart with the admittance chart, i.e. the impedance chart after rotation through 180°. This is shown in Fig. 3.4. In order to achieve input matching, we move from R_0 at the centre of the chart, to I_1 by adding a shunt capacitance such that $jR_0C_1\omega = 2.25j$ (read off on the admittance chart), and we indeed find that $C_1\omega = 45$ mS. We then go from I_1 to the point representing Γ_1 by adding a series inductance such that

$$\frac{jL_1\omega}{R_0} = 0 - (-0.37j) \qquad \text{or} \qquad L_1\omega = 18.5\,\Omega$$

Similarly, in order to achieve output matching, we go from the point representing R_0 to I_2 by adding a series capacitance such that

$$-\frac{j}{R_0C_2\omega} = -2.3j \qquad \text{or} \qquad C_2\omega = 8.7\,\text{mS}$$

Then, on the admittance chart, we go from I_2 to the point representing Γ_2 by adding a shunt inductance such that

$$-j\frac{R_0}{L_2\omega} = -0.185j - 0.365j$$

and hence

$$L_2\omega = 90\,\Omega$$

In any case, both these methods lead to the block diagram shown in Fig. 3.5. Although we end up with orders of magnitude of a few picofarads for the capacitances and only a few nanohenries for the inductances, it is still possible to use lumped components. In fact, adjustable or fixed capacitors of 1 pF are readily available commercially for frequencies of up to, say, 1 or 2 GHz. Miniature

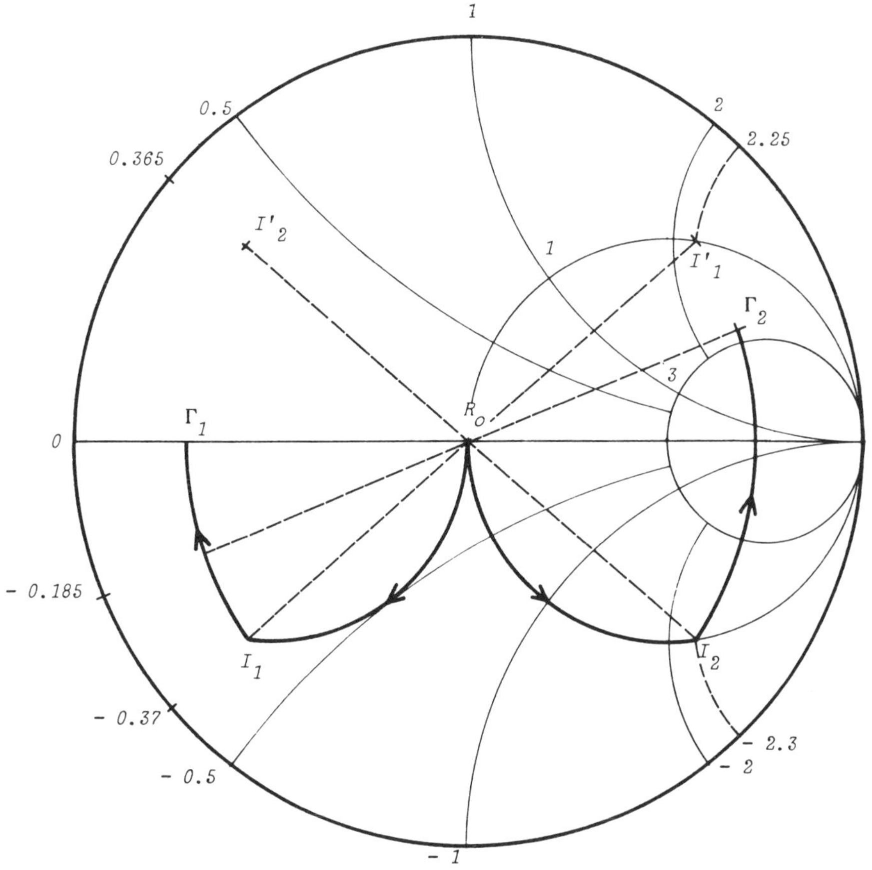

Fig. 3.4 *Graphical synthesis of matching circuits.*

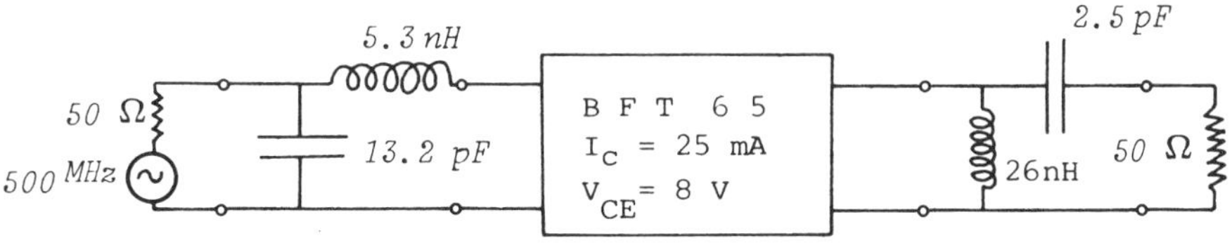

Fig. 3.5 *500 MHz amplifier: the numerical values are those required for input and output matching.*

inductors in the form of moulded coils are also available. However, they can easily be fabricated in Cu–Ag wire, and pre-characterization can be carried out using the practical formula

$$L = \frac{100d^2}{4d + 11l} n^2 \tag{3.5}$$

where L is in nanohenries, n is the number of turns, and d and l expressed in centimetres are the diameter and length respectively of the coil. The overvoltage coefficient is maximum for d/l ratio of 2.5. A third method involves drawing circular or square spirals on a printed circuit; double-sided technology can reduce the internal connection problem.

At these frequencies, it is necessary to employ a ground plane correctly connected to the enclosure; this also makes it possible to provide connections to the input and output sockets using 50 Ω lines. The power supply, carefully isolated, is of course connected through this ground plane. Provided that such precautions are taken, the risks of unwanted oscillations are eliminated. A design has yielded the gain curve shown in Fig. 3.6.

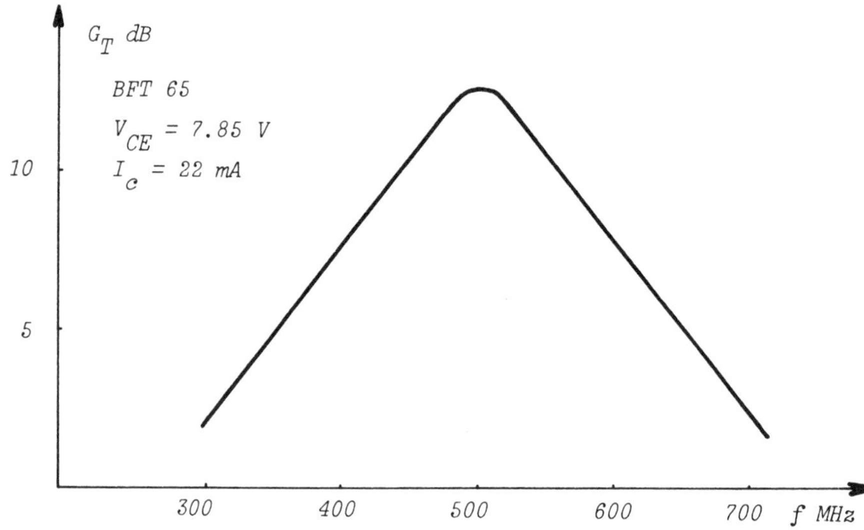

Fig. 3.6 *Single-stage narrow band amplifier.*

Now suppose that we wish to design a two-stage amplifier with each stage identical with that analysed above. The configuration can be as shown in Fig. 3.7. We still have to design an interstage matching circuit, and the Smith chart proves to be a very useful tool for this purpose.

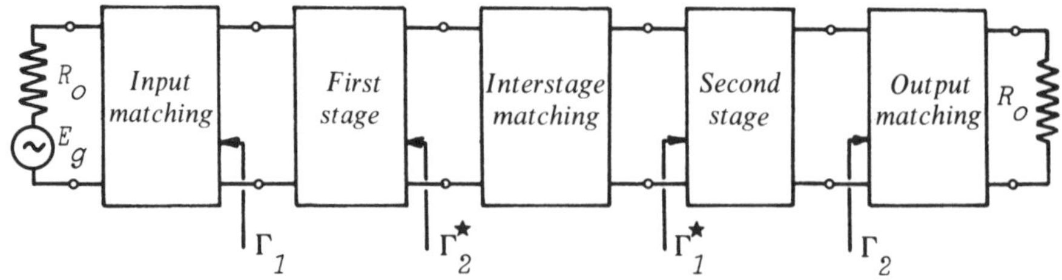

Fig. 3.7 *Two-stage narrow band amplifier.*

We have seen that the input circuit of the first stage presents an impedance Z_1, or Γ_1 in reflection coefficient terms. Therefore the reflection coefficient at the output of the first stage is

$$\Gamma_2^* = 0.7386\,(-23°)$$

Similarly, the output matching circuit presents a reflection coefficient Γ_2 to its input. Thus the second stage will provide a reflection coefficient

$$\Gamma_1^* = 0.7213\,(-180°)$$

The interstage matching circuit therefore 'sees' the reflection coefficient Γ_1^* of the second stage which it must transform into Γ_2 at its input. We shall therefore plot the points representing Γ_1^* and Γ_2 on the Smith chart (Fig. 3.8) We can reach Γ_2 from Γ_1^* by two paths.

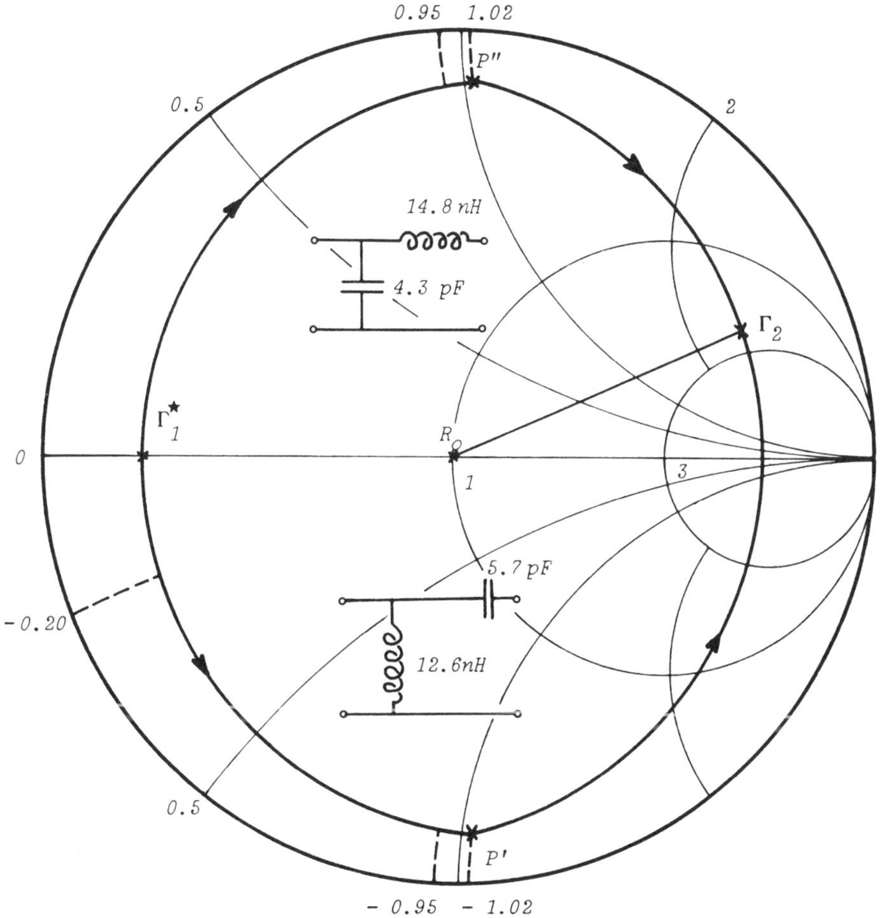

Fig. 3.8 Interstage matching.

On the lower path, moving on the constant-resistance circle, we arrive at P′ by adding a series capacitance C' such that

$$-\frac{j}{R_0 C' \omega} = -1.02j - 0.0j$$

Then at 550 MHz we obtain $C' = 5.7$ pF. Now considering the admittance chart, we reach Γ_2 by moving round the constant-conductance circle, i.e. by adding a shunt self-inductance L' such that

$$-\frac{jR_0}{L'\omega} = -0.20j - 0.95j$$

Hence $L' = 12.6$ nH.

On the upper path we reach P″ by adding a series self-inductance L'' such that

$$\frac{jL''\omega}{R_0} = 1.02j - 0.0j$$

Hence $L'' = 14.8$ nH. The point representing Γ_2 is then reached by adding a shunt capacitance C'' such that

$$jR_0 C''\omega = -0.20j - (-0.95j)$$

Hence $C'' = 4.3$ pF.

Two configurations are therefore possible as shown in Fig. 3.8; in practice, we select the one whose components can most easily be realized.

Note

We could have looked from the input side and assumed that the interstage circuit, 'seeing' Γ_2^*, should present a reflection coefficient Γ_1 at the output. On the Smith chart, this amounts to moving from the point representing Γ_2^* to that representing Γ_1, i.e. producing a symmetry about the zero reactance axis. Consequently, the lower path would have given the upper circuit and the upper path the lower circuit, and we therefore obtain the same (two) solutions.

3.2. Wide band amplifier

This single-stage amplifier must have a transducer power gain which is as constant as possible in the band of frequencies to be amplified between 500 and 900 MHz. The transistor selected is the BFR 90 (silicon n–p–n) for which the S parameters at various frequencies at the DC operating point $\{V_{CE} = 8\text{ V};\ I_c = 14\text{ mA}\}$ have the values given in Table 3.1. The calculations are performed graphically using a Smith chart.

In the case of the four frequencies in Table 3.1 it is useful to plot initially the critical instability circles C_2 and C_1 at the input and output respectively which

Table 3.1

Frequency (MHz)	S_{11e}	S_{22e}	S_{21e}	S_{12e}
300	0.267 (−88.5°)	0.57 (−30.8°)	11.61 (112°)	0.043 (68°)
500	0.178 (−122°)	0.52 (−29°)	7.37 (99°)	0.060 (70°)
700	0.122 (−156°)	0.49 (−28°)	5.45 (89°)	0.082 (69°)
900	0.111 (−179°)	0.46 (−27°)	3.75 (77°)	0.106 (69°)

will determine, as and when required, the forbidden regions for the load impedance (reflection coefficient Γ_2) and the source impedance (reflection coefficient Γ_1).

Table 3.2 was compiled using Eqns (2.20) and (2.21) to determine the radii R_i and the centres Ω_i of circles C_i ($i = 1, 2$). It follows directly from this table that the circles C_1 contain the centre of the chart since $|\Delta|$ is greater than $|S_{11}|$ and that, consequently, the area inside them is a stable region ($|S_{22}| < 1$). The reverse is true of the circles C_2, given that this time $|\Delta|$ is less than $|S_{22}|$.

Table 3.2

Frequency (MHz)	Determinant		Critical instability circles			
			C_1 at output		C_2 at input	
	Δ	$\|\Delta\|^2$	R_1	Ω_1	R_2	Ω_2
300	0.44500 (−17.35°)	0.1980	3.940	3.19282 (−53.7°)	3.934	4.77429 (41.86°)
500	0.37603 (−20.10°)	0.1414	4.030	3.09751 (−32.2°)	3.428	4.38829 (34.13°)
700	0.39048 (−24.71°)	0.1525	3.248	2.24195 (−11.3°)	5.100	6.103915 (29.80°)
900	0.34701 (−35.17°)	0.1204	3.678	2.49619 (4.4°)	4.359	5.46122 (27.71°)

The elements calculated above have been used to draw the instability circles on a Smith chart (Fig. 3.9). We found that an unstable zone occurs at frequencies below 700 MHz, and that this effect became more marked with decreasing frequency. Although in the remainder of this analysis we shall use the unilateral transistor model which, in essence, is unconditionally stable, these forbidden regions corresponding to the real transistor must be taken into consideration when both the input and the output reflection coefficients are selected.

We have calculated, as a function of frequency, the maximum transducer power gain $G_{TU(max)}$ for the transistor which is assumed to be unilateral. $G_{TU(max)}$ is the sum of three contributions from the input, the transistor and the output. This gain is compared, for $|K| > 1$, with $G_{T(max)}$ (see Eqns (2.14) and (2.17)) in Table 3.3. The difference between $G_{T(max)}$ and $G_{TU(max)}$ decreases with increasing frequency from 1.85 dB at 700 MHz to only 0.75 dB at 900 MHz, which is interesting since

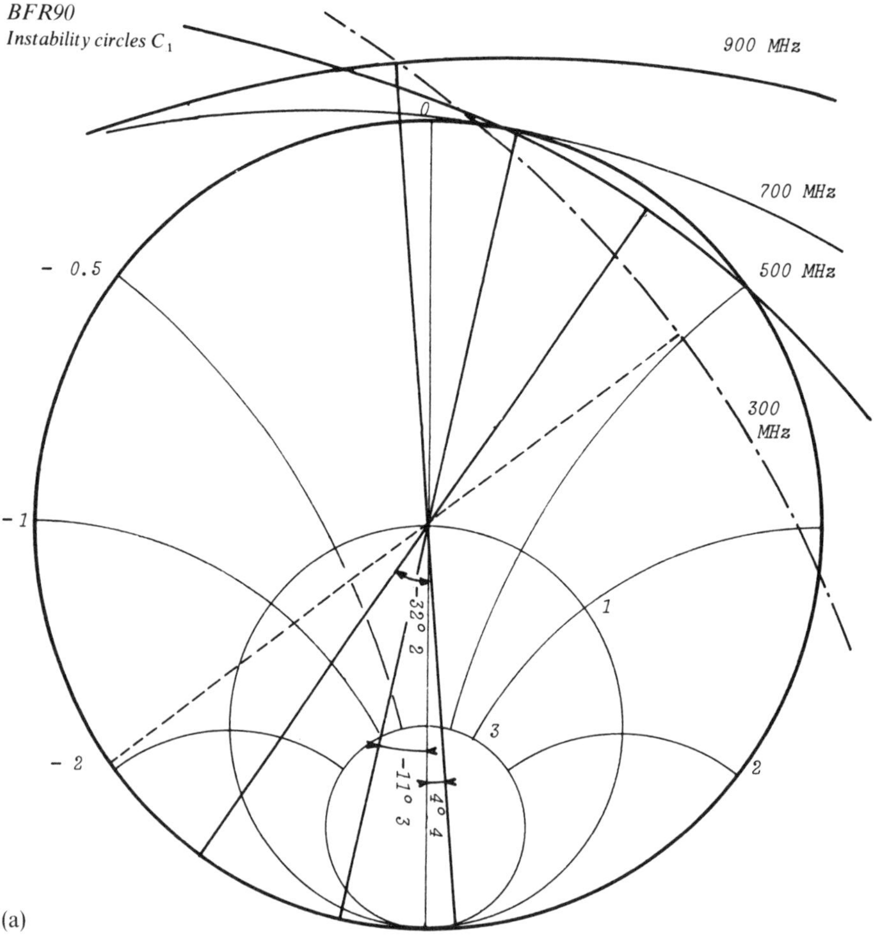

(a)

Fig. 3.9 *(a) Loci of the reflection coefficients Γ_1 presented by the source which make the output critical; (b) loci of the reflection coefficients Γ_2 presented by the load which make the input critical.*

Table 3.3

Frequency (MHz)	Stability coefficient K	Gain $G_{T(max)}$ at matching (dB)	Input gain $1/(1-\|S_{11}\|^2)$ (dB)	Transistor gain S_{21} (dB)	Output gain $1/(1-\|S_{22}\|^2)$ (dB)	Maximum unilateral gain $G_{TU(max)}$ (dB)
300	0.8030		0.32	21.30	1.70	23.32
500	0.9490		0.14	17.35	1.37	18.86
700	1.0041	17.83	0.065	14.73	1.19	15.985
900	1.1277	13.315	0.055	11.48	1.03	12.57

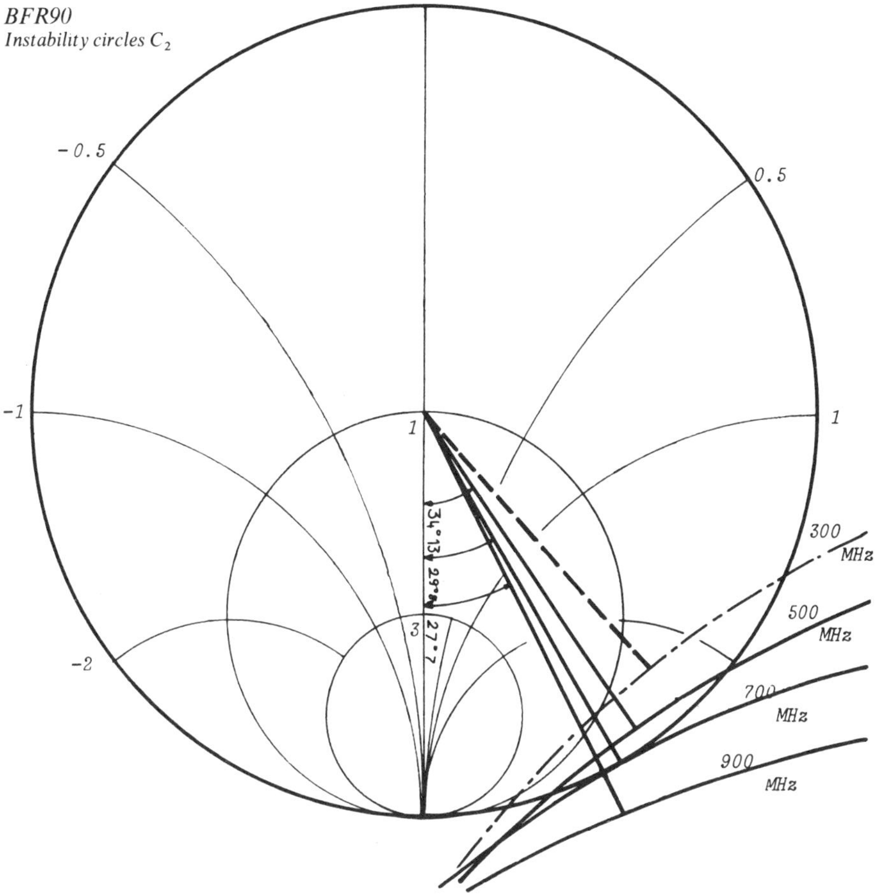

Fig. 3.9 (*b*)

we are attempting to obtain the maximum gain of the stage. Moreover, the gain provided by input matching, i.e. for $\Gamma_1 = S_{11}^*$, which gives

$$G_{1\max} = \frac{1}{1-|S_{11}|^2}$$

in accordance with Eqn. (2.32b), is particularly small.

It will not therefore be necessary *a priori* to provide an input matching circuit; we shall then have

$$\Gamma_1 = 0 \qquad G_1 = 0\,\text{dB}$$

The advantage of a transistor for which $|S_{11}|$ is low should be noted. Thus in this case we have

$$G_{TU} = 0 + 11.48 + 1.03 \approx 12.5\,\text{dB} \qquad \text{at } 900\text{ MHz}$$

$$G_{TU} = 0 + 17.35 + 1.37 \approx 18.7\,\text{dB} \qquad \text{at } 500\text{ MHz}$$

As we are attempting to obtain a constant gain in the 500–900 MHz band, it is necessary to mismatch the output at the lower frequencies in order to make the gain at 500 MHz fall by $18.7-12.5 = 6.2$ dB; the 'gain' due to the output circuit at this frequency must therefore be

$$G_2 = 1.37 - 6.2 \approx 4.8 \text{ dB}$$

Therefore the problem is as follows: to design an output matching circuit which, at 500 MHz, will present a reflection coefficient $\Gamma_2(500)$ such that $G_2(500) = -4.8$ dB and which, at 900 MHz, will present $\Gamma_2(900)$ such that $G_2(900) \approx 1.03$ dB.

We have seen that the locus of the reflection coefficients Γ_2 such that G_2 is constant is a circle. In order to characterize the circuit graphically, we shall define the circles corresponding to $G_2(500)$ values of -3, -4 and -5 dB, and to $G_2(900)$ values of 0.5 and 0.8 dB. The components of these circles, radii R_{G_2} and centres Ω_{G_2}, are given in Table 3.4 which was computed using Eqn (2.35) with subscript 2 replacing subscript 1. These circles are plotted on a Smith chart in Fig. 3.10. We now proceed by successive approximation; we have adopted the following paths.

Table 3.4

G_2 (dB)	R_{G_2}	Ω_{G_2}
Frequency 500 MHz		
−3	0.7014	0.2295 (29°)
−4	0.7605	0.1868 (29°)
−5	0.8081	0.1514 (29°)
Frequency 900 MHz		
+0.5	0.2745	0.4171 (27°)
+0.8	0.1824	0.4408 (27°)

(1) Starting at 900 MHz, we move from O to P′ by adding a series capacitor whose capacitance C is obtained from the expression

$$-\frac{\mathrm{j}}{R_0 C\omega} = -1.73\mathrm{j} - 0.0\mathrm{j}$$

Hence

$$C = 2.05 \text{ pF}$$

(2) Now at 500 MHz this capacitance value takes us to point P such that

$$-\frac{\mathrm{j}}{R_0 C\omega} = -\mathrm{j}x$$

Thus

$$x = 3.11$$

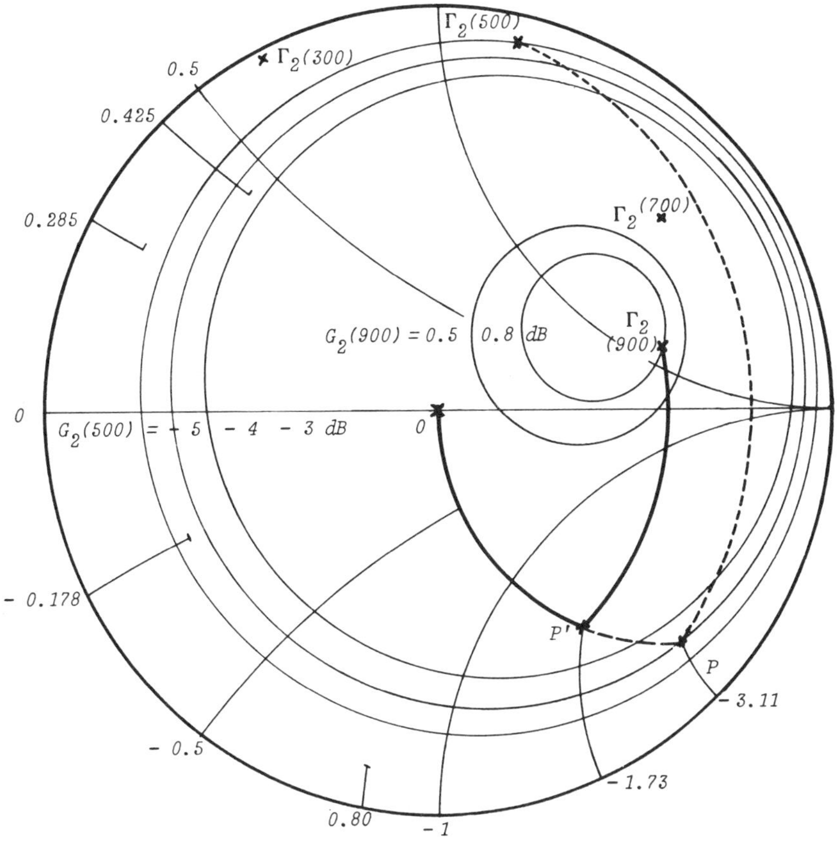

Fig. 3:10 *Wide band output matching of the amplifier.*

(3) Still at 500 MHz, but this time moving on the admittance chart, we rejoin the circle $G_2 = -4.8$ dB (actually -5 dB on the figure) at the point

$$\Gamma_2(500) = 0.902\,(77.5^\circ)$$

To do this, it is necessary to add a parallel inductor whose self-inductance L is defined by

$$-\frac{jR_0}{L\omega} = -0.80j - 0.285j$$

Hence

$$L = 14.67\ \text{nH}$$

(4) We return to 900 MHz, and this inductance value finally brings us to the point

$$\Gamma_2(900) = 0.613\,(21^\circ)$$

This value follows from the relation

$$-\frac{jR_0}{L\omega} = -jy - 0.425j$$

which gives $y = 0.178$.

For a load $R_0 = 50\,\Omega$, the wide band output matching circuit will therefore simply comprise a series capacitor followed by a shunt inductor, as shown in Fig. 3.11. This circuit has been fabricated, and Fig. 3.12 shows the transducer power gain obtained in the band 500–900 MHz.

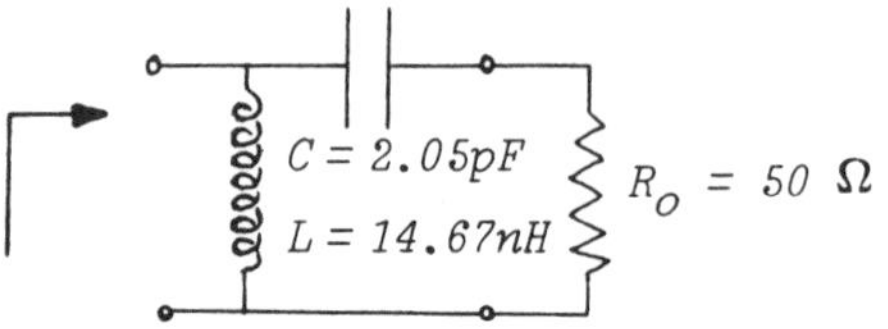

Fig. 3.11 *Output circuit for wide band matching:* $\Gamma_2(500) = 0.902\ (77.5°)$; $\Gamma_2(900) = 0.613\ (21°)$

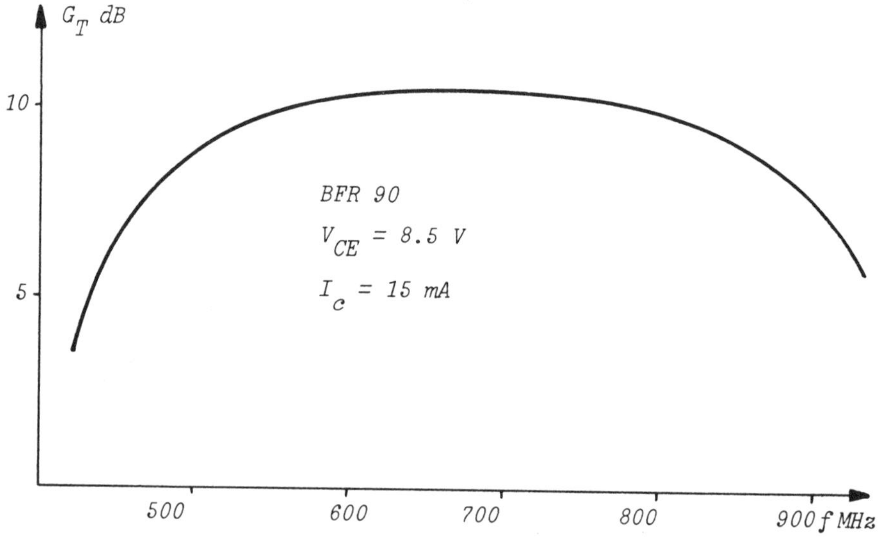

Fig. 3.12 *Single-stage wide band amplifier.*

3.3. Low-noise preamplifier using distributed-element circuits

We propose to design a preamplifier operating at 2.5 GHz using a low-noise n–p–n bipolar transistor (HXTR 6102). This device, adapted to suit microstrip technology, is shown schematically in Fig. 3.13. The S_{11} and S_{22} parameters for the operating point $\{V_{CE} = 10\,V; I_c = 4\,mA\}$ are shown on a Smith chart (Fig. 3.14) for frequencies varying from a few hundred megahertz to 7 GHz. The amplitude of S_{21} in decibels and its phase in degrees are plotted in Fig. 3.15 using a logarithmic frequency scale. A decrease of about 6 dB per octave is observed. With regard to the noise parameters, we have plotted on a Smith chart the locus of the reflection coefficients Γ_{opt} giving the minimum noise factor F_{min} as a function of frequency (Fig. 3.16). The values of Γ_{opt} can be read off on the circumference of the chart. We have also plotted the variations in F_{min} and the associated noise resistance R_n, again as a function of frequency (Fig. 3.17).

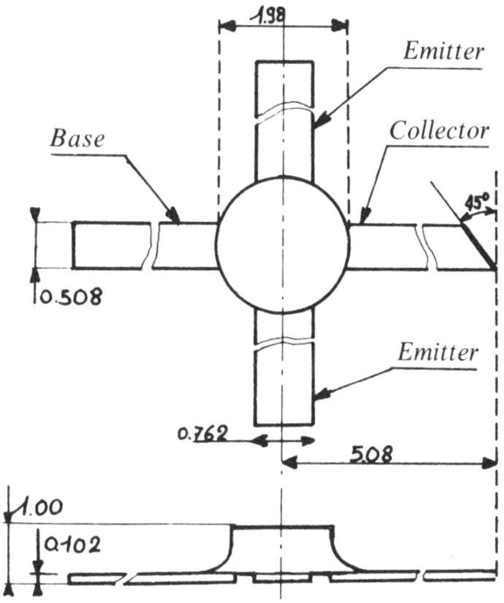

Fig. 3.13 *N–p–n transistor HXTR 6102 (dimensions in millimetres).*

Thus we shall adopt the following values at 2.5 GHz: for the scattering parameters

$$S_{11} = 0.570\,(-163^\circ)$$
$$S_{22} = 0.760\,(-50^\circ)$$
$$S_{21} = 2.646\,(59^\circ)$$
$$S_{12} = 0.042\,(25^\circ)$$

and for the noise parameters

$$F_{min} = 1.8\,\text{dB}$$
$$R_n = 13.8\,\Omega$$
$$\Gamma_{opt} = 0.412\,(104^\circ)$$

From the first parameter set we obtain

$$\Delta = 0.3953\,(161.5^\circ)$$
$$K = 1.1417$$

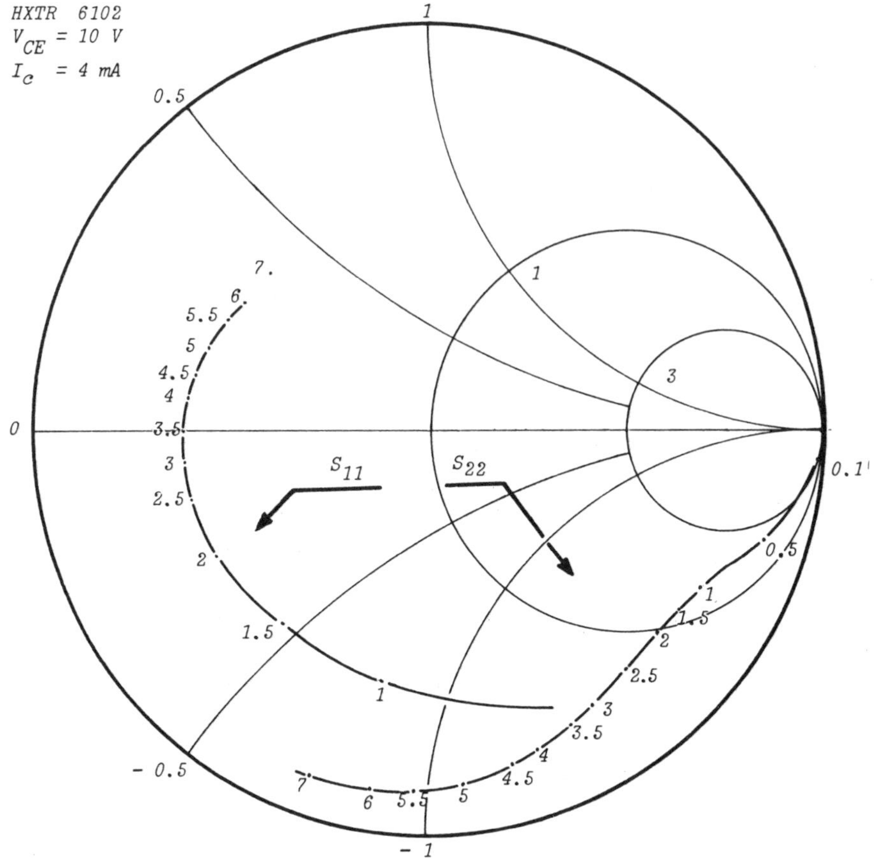

Fig. 3.14 *Reflection coefficients S_{11} and S_{22} as a function of frequency in gigahertz.*

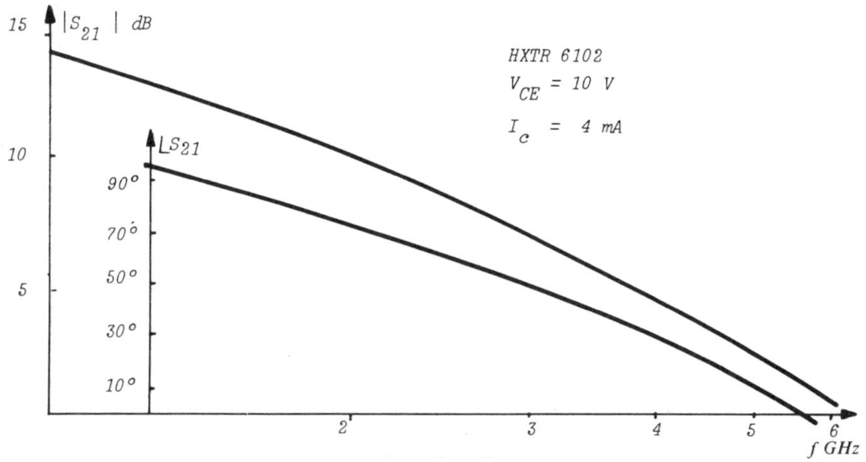

Fig. 3.15 *Transmission coefficient S_{21} as a function of frequency.*

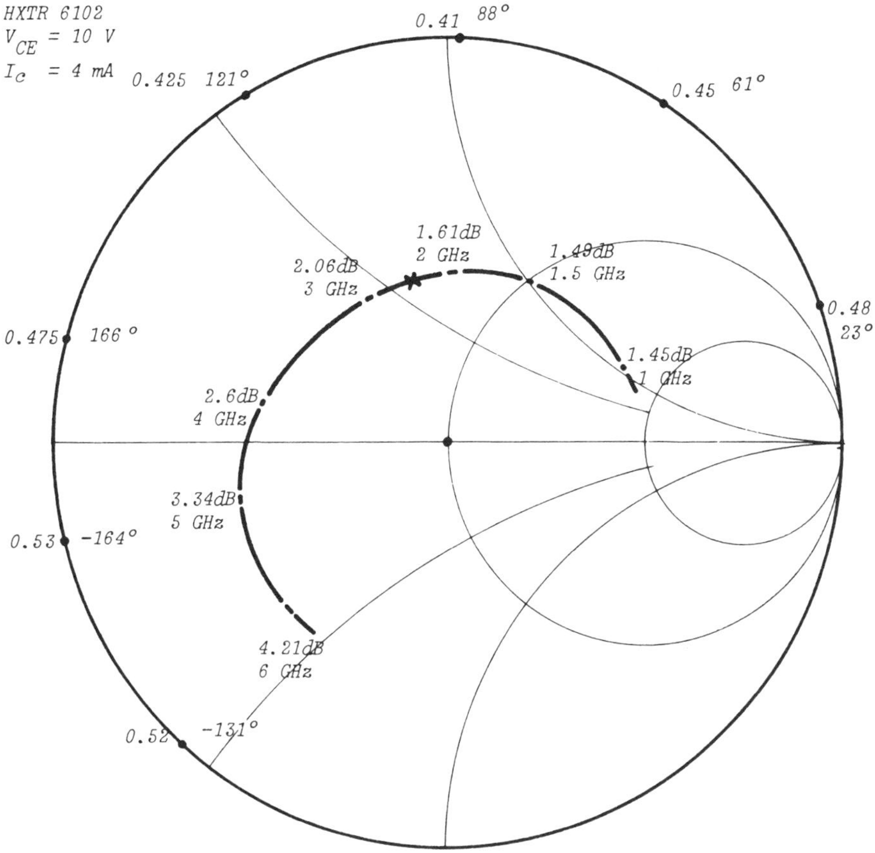

Fig. 3.16 *Variations in the reflection coefficient Γ_{opt} of the source giving the minimum noise factor F_{min} as a function of frequency.*

(see Eqn (2.14a)). Therefore the transistor is unconditionally stable, and in this case the pair of reflection coefficients allowing simultaneous input–output matching is $\{\Gamma_1', \Gamma_2'\}$. We recall that these are the coefficients which the transistor must see at the input and output respectively. Using Eqns (2.12) and (2.13) we calculate

$$B_1 = 0.5910 \qquad C_1 = 0.28905\,(182^\circ)$$

and also

$$B_2 = 1.0965 \qquad C_2 = 0.5448\,(-56^\circ)$$

We can show that

$$(B_1{}^2 - 4|C_1|^2)^{\frac{1}{2}} = (B_2{}^2 - 4|C_2|^2)^{\frac{1}{2}} = 0.1225$$

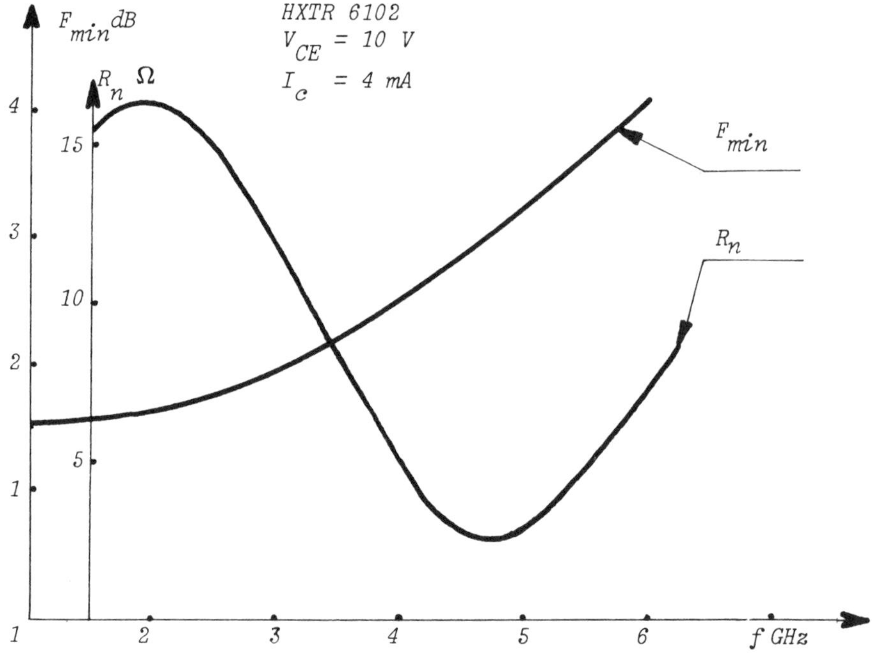

Fig. 3.17 *Variations in the minimum noise factor F_{min} and the associated resistance R_n as a function of frequency.*

and hence finally

$$\Gamma_1' = 0.810(-182°)$$

$$\Gamma_2' = 0.894(56°)$$

If we adopted this pair of values, the transducer power gain would be maximum. Using Eqn (2.17) we obtain

$$G_{T(max)} = 15.7\,\text{dB}$$

However, as Γ_1' is different from Γ_{opt}, the noise factor is increased and from Eqn (2.54) we find

$$F = 5.6\,\text{dB}$$

We could equally well have plotted the constant-noise-factor circles in accordance with (2.56); this has been done in Fig. 3.18 where we find that the point representing Γ_1' lies approximately on the circle $F = 5.5$ dB.

In conclusion, in the present case simultaneous matching is not compatible with a low noise factor.

We now consider the unilateral transistor model. In this case, matching is achieved not by presenting Γ_1' and Γ_2', but by presenting $S_{11}{}^*$ and $S_{22}{}^*$ respectively. The transducer power gain is then maximum. It is expressed in the

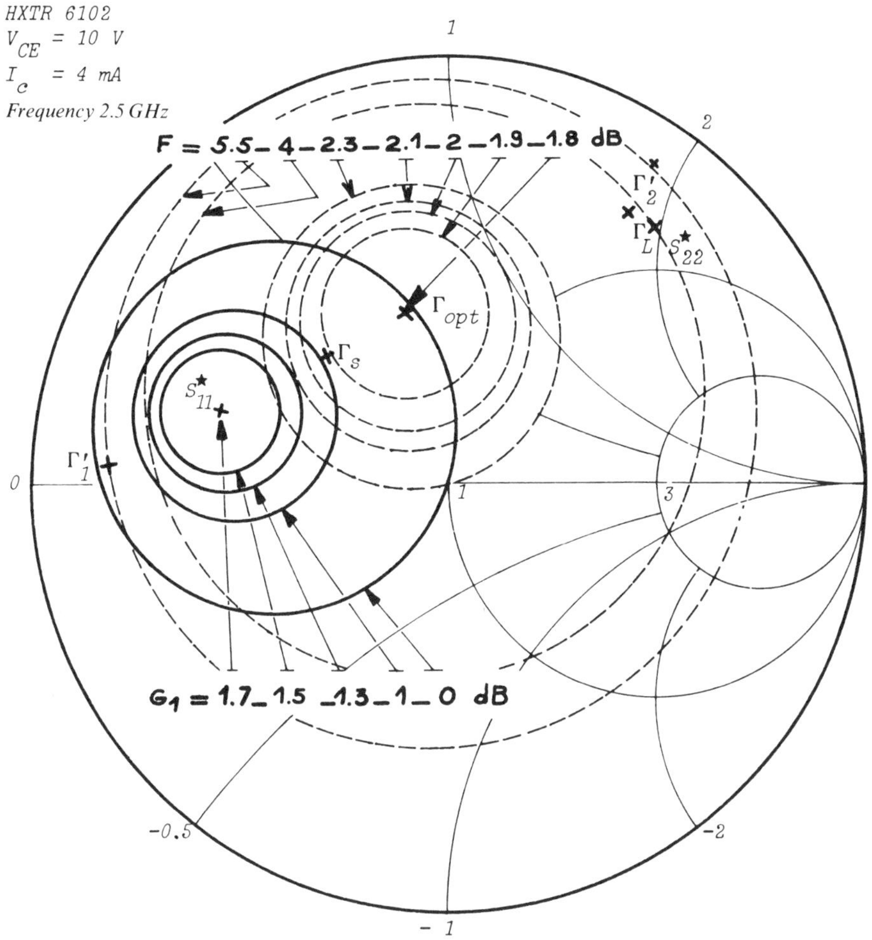

Fig. 3.18 Use of the circles method to find the reflection coefficient Γ_s providing a satisfactory compromise between gain and noise factor.

form of the product of three terms in accordance with Eqn (2.32a) such that

$$G_{TU(max)} = 1.7\,dB + 8.45\,dB + 3.75\,dB$$

or

$$G_{TU(max)} = 13.9\,dB$$

The difference with respect to $G_{T(max)}$ is 1.8 dB, which is not negligible; however, it is interesting to note the dominant contribution of the output matching and the fact that Γ_2' is not very different from S_{22}^*.

Still assuming that the transistor is unilateral, we plot on Fig. 3.18 circles of constant gain G_1 at the input; for this we use Eqns (2.35). It appears inadvisable to select a source impedance Z_s such that the reflection coefficient Γ_s is equal to Γ_{opt}, because then G_1 is roughly equal to 0 dB, i.e. there is a loss of 1.7 dB with respect to the optimum $\Gamma_s = S_{11}{}^*$. We obtain a good compromise by selecting the tangential point of the circle:

$$G_1 = 1\,\text{dB} \qquad F = 1.9\,\text{dB}$$

We shall therefore take

$$\Gamma_s = 0.395\,(129°)$$

We can use this reflection coefficient to calculate the noise factor more precisely (see Eqn (2.54)) and find that

$$F = 1.92\,\text{dB}$$

Let us now leave this unilateral transistor model which we have used merely to find a satisfactory compromise between an adequate gain and a noise factor which is not excessively high. We therefore have to match the output of the real transistor, i.e. to calculate the reflection coefficient Γ_L which the load must present. Using relation (2.11) we derive the equalities

$$\Gamma_L = S_{22}{}^* + \frac{S_{12}{}^* S_{21}{}^*}{1/\Gamma_s{}^* - S_{11}{}^*} = S_{22}{}^* + \delta S_{22}{}^*$$

The correction term is

$$\delta S_{22}{}^* = -0.04865 + 0.02186\text{j}$$

which should be compared with

$$S_{22}{}^* = 0.48852 + 0.58219\text{j}$$

In modulus–argument form we therefore have

$$\Gamma_L = 0.747\,(54°)$$

Since we know the pair $\{\Gamma_S;\ \Gamma_L\}$, we are able to calculate the theoretical transducer power gain. We use the formula derived from Eqn (2.6), i.e.

$$G_T = \frac{1-|\Gamma_s|^2}{|1-\Gamma_s S_{11}|^2}|S_{21}|^2 \frac{1-|\Gamma_L|^2}{|1-\Gamma_L S'_{22}|^2}$$

where

$$S'_{22} = S_{22} + \frac{S_{12} S_{21}}{1/\Gamma_s - S_{11}}$$

In the present case, since the output is matched, we note that

$$S'_{22} = \Gamma_L{}^*$$

and hence

$$G'_2 = \frac{1-|\Gamma_L|^2}{|1-\Gamma_L S'_{22}|^2} = \frac{1}{1-|\Gamma_L|^2}$$

Therefore G_T represents an available power gain which we shall denote by G_A; we shall return to this point in greater detail later on. Numerically, we have

$$10\log\left(\frac{1-|\Gamma_s|^2}{|1-\Gamma_s S_{11}|^2}\right) = 0.95\,\text{dB}$$

$$20\log|S_{21}| = 8.45\,\text{dB}$$

$$10\log\left(\frac{1}{1-|\Gamma_L|^2}\right) = 3.55\,\text{dB} \qquad (G'_2\ \text{max})$$

and hence

$$G_A = 12.95\,\text{dB}$$

Note 1

An important characteristic of the amplifier is its noise figure (Eqn (2.58)) which is given here by

$$M = 1 + \frac{F-1}{1-1/G_A}$$

so that

$$M = 2\,\text{dB}$$

Simultaneous matching would have given $M = 5.7\,\text{dB}$!

Note 2

The expression for G_{TU} given in Eqn (2.28) in the form of a product of three terms G_1, G_0 and G_2 and that used above for G_T have the same morphology; moreover, the first two terms are identical. Selecting Γ_s gives the exact value $G_1 = 0.95\,\text{dB}$. However,

$$G_2 = 10\log\left(\frac{1-|\Gamma_L|^2}{|1-\Gamma_L S_{22}|^2}\right) = 3.67\,\text{dB}$$

which should be compared with

$$G'_{2(\text{max})} = 3.55\,\text{dB}$$

We therefore obtain

$$G_{TU} = 13.07\,\text{dB}$$

If we use inequality (2.39) in which $u = 0.0915$ in the present case, we obtain

$$12.31\,\text{dB} < G_A < 13.90\,\text{dB}$$

Given the size of the spread thus obtained, this inequality is unlikely to be of practical use. It is preferable to calculate G_T directly, which is no more tedious provided that we select the most convenient formulation.

The next step consists of constructing matching circuits which, from $R_0 = 50\,\Omega$, present Γ_s at the input and Γ_L at the output. At this frequency we select microstrip technology based on the structure shown in Fig. 3.19.

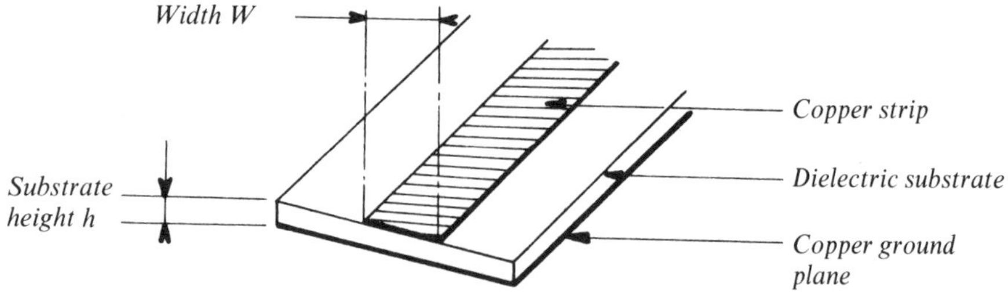

Fig. 3.19 *Structure of a microstrip line.*

The attenuation α (in nepers per metre) introduced by such lines is the sum of the attenuation α_c in the conductor and the attenuation α_d in the dielectric. It is useful to characterize a line by the product $\alpha\lambda_m$ (in decibels) where λ_m is the wavelength in the microstrip structure at the operating frequency. The practical form of the general expression for $Q = \beta/2\alpha$ is therefore

$$Q = \frac{\pi}{\alpha\lambda_m} \tag{3.6}$$

since $\beta = 2\pi/\lambda_m$.

It is necessary to check that the 'Q', which is usually measured by resonance methods, is sufficient (at least about 100 in the frequency range used here).

We now need to know two fundamental quantities for constructing the matching circuits: the characteristic impedance Z_c and the normalized wavelength λ_m/λ_0 (λ_0 is the free-space wavelength associated with the frequency f). These quantities depend on the geometry of the line (in fact to a first approximation only the ratio W/h is involved) and the relative permittivity ε_r of the dielectric. The literature contains more or less empirical formulae (which are quite accurate, however) giving Z_c and λ_m/λ_0; we shall use

$$Z_c = \frac{377}{\varepsilon_r^{\frac{1}{2}}}\frac{h}{W}\frac{1}{1+1.735\varepsilon_r^{-0.0724}(W/h)^{-0.836}} \tag{3.7}$$

$$\lambda_m = \frac{\lambda_0}{\varepsilon_r^{\frac{1}{2}}}\left\{\frac{\varepsilon_r}{1+0.63(\varepsilon_r-1)(W/h)^{0.1255}}\right\}^{\frac{1}{2}} \tag{3.8a}$$

for $W/h \geqslant 0.6$ and

$$\lambda_m = \frac{\lambda_0}{\varepsilon_r^{\frac{1}{2}}}\left\{\frac{\varepsilon_r}{1+0.6(\varepsilon_r-1)(W/h)^{0.0297}}\right\}^{\frac{1}{2}} \tag{3.8b}$$

for $W/h \leqslant 0.6$, where W is the width of the line and h is the height of the substrate.

The *Microwave Engineers' Handbook* (see bibliographical references) gives the ratio λ_0/λ_m and the characteristic impedance Z_c as a function of the relative width W/h of the line for various values of the relative dielectric constant ε_r. Figures 3.20 and 3.21, taken from the work of Wheeler, reproduce the relevant charts. As we can see, there are two sets of tables depending on whether the line is narrow ($W/h < 1$) or wide ($W/h > 1$).

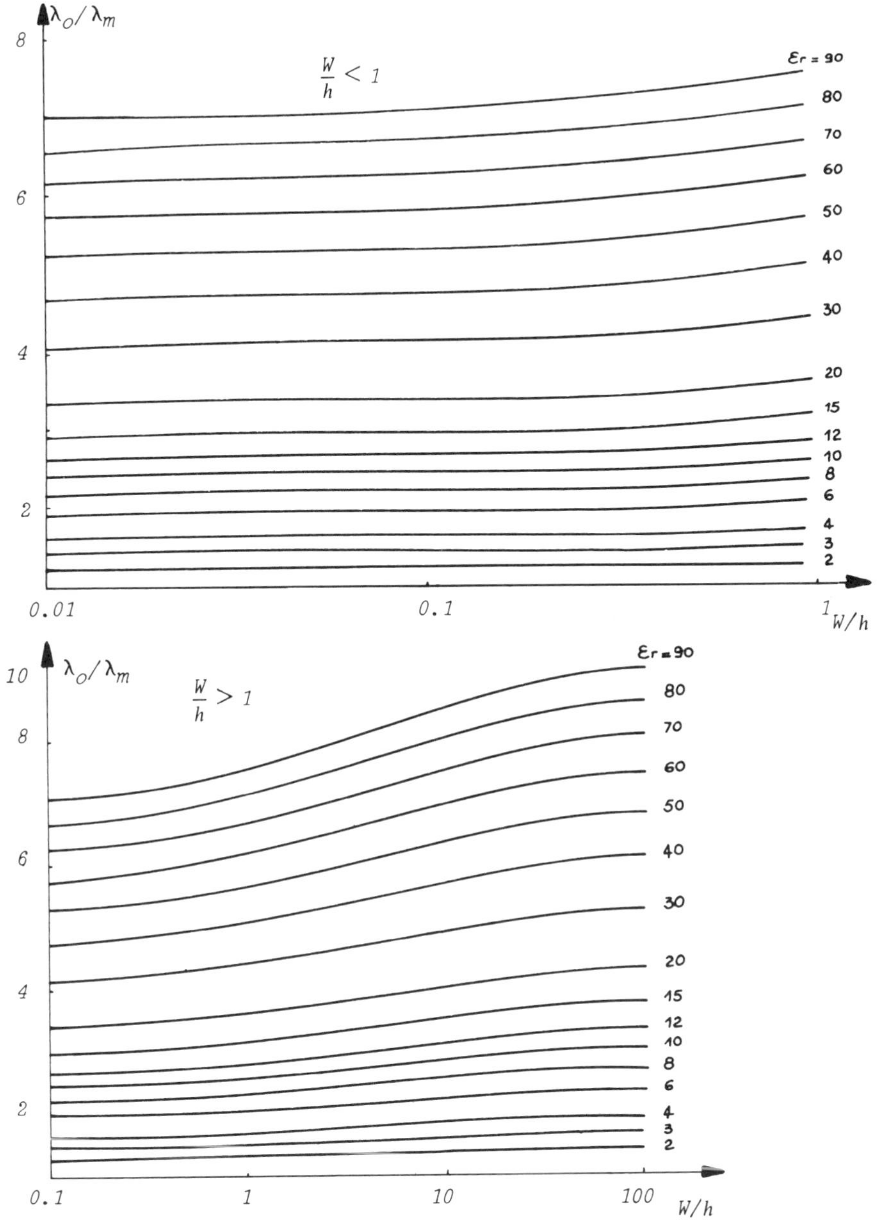

Fig. 3.20 *Changes in the ratio λ_0/λ_m as a function of the geometric characteristic W/h of the line for various values of the relative permittivity ε_r. (After Wheeler.)*

The stubs used in the matching circuits were specified as follows. Given Z_c, we obtained W/h using the charts; we were then able to calculate λ_m, this time using Eqns (3.8a) or (3.8b).

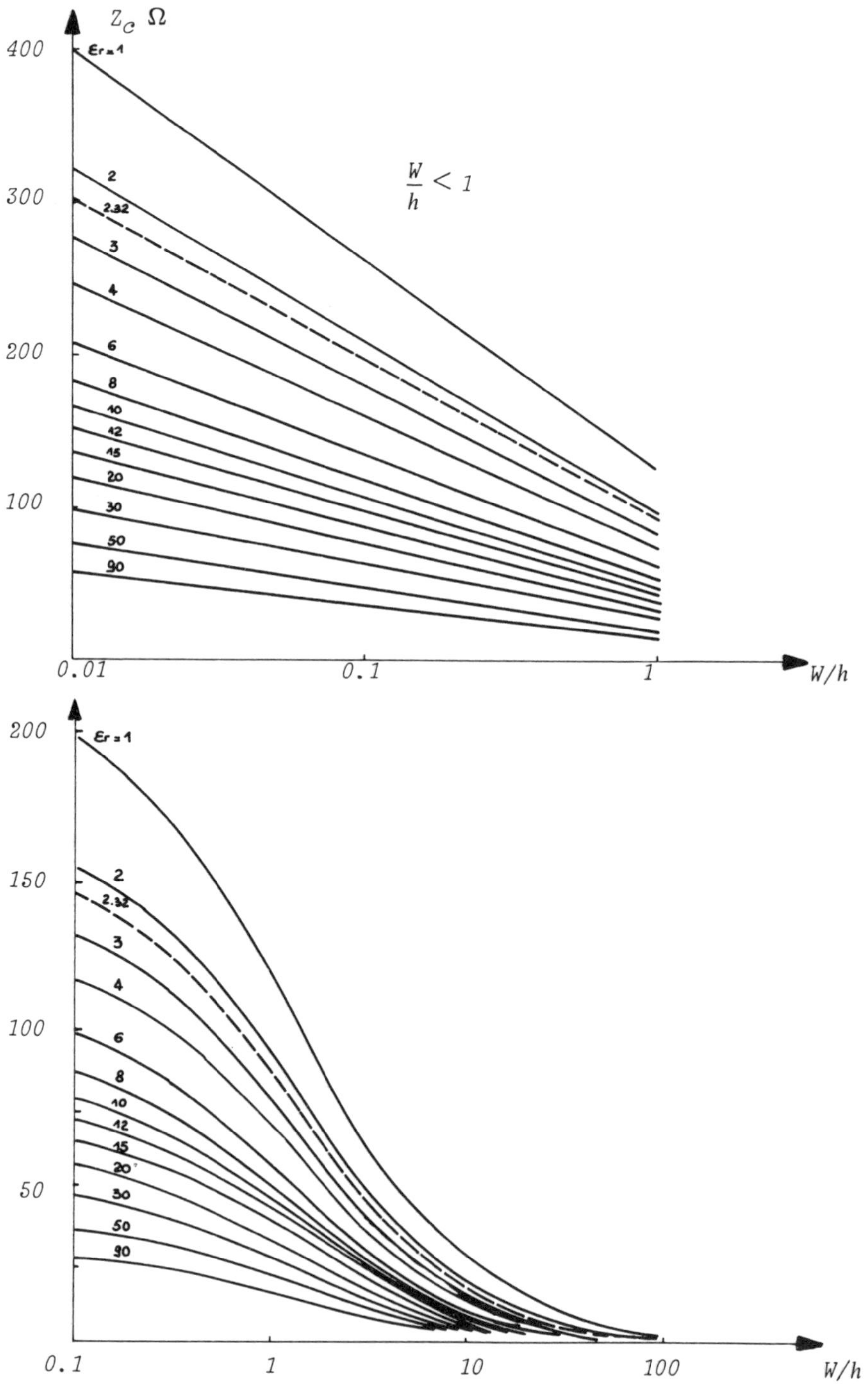

Fig.3.21 *Changes in the impedance Z_c of the line as a function of its geometric characteristic W/h for various values of the relative permittivity ε_r. (After Wheeler.)*

It should be added that the widespread use of alumina-type substrates at frequencies above a few gigahertz has meant that other formulae have been worked out giving the ratios W/h and λ_m/λ_0 as functions of Z_c and ε_r with a remarkable degree of accuracy (better than 1%) for $\varepsilon_r \approx 10$. These formulae can be found in Appendix B.

A last point to be discussed before going on to numerical applications is the determination of the lengths l of the lines used; in the present case, we shall restrict ourselves to the values $l = \lambda_m/4$ and $l = \lambda_m/8$, but this certainly cannot be taken as a general rule. In fact, we know that, with an impedance Z_T, a length l of line of characteristic impedance Z_c will present an impedance of $Z(l)$ such that

$$Z(l) = \frac{Z_T + jZ_c \tan\{2\pi(l/\lambda_m)\}}{1 + j(Z_T/Z_c)\tan\{2\pi(l/\lambda_m)\}} \tag{3.9}$$

In order to derive this formula, it is merely necessary to return to the definition of the impedance $Z(l)$ which we expand using (1.52) and (1.53):

$$Z(l) = \frac{V(l)}{I(l)} = Z_c \frac{V_T^{+} e^{\gamma l} + V_T^{-} e^{-\gamma l}}{V_T^{+} e^{\gamma l} - V_T^{-} e^{-\gamma l}}$$

We therefore divide top and bottom by V_T^{+} in order to obtain the reflection coefficient in T:

$$\frac{V_T^{-}}{V_T^{+}} = \frac{Z_T - Z_c}{Z_T + Z_c}$$

It then readily follows that

$$Z(l) = Z_c \frac{Z_T(e^{\gamma l} + e^{-\gamma l}) + Z_c(e^{\gamma l} - e^{-\gamma l})}{Z_T(e^{\gamma l} - e^{-\gamma l}) + Z_c(e^{\gamma l} + e^{-\gamma l})}$$

Hence

$$Z(l) = Z_c \frac{Z_T + Z_c \tanh(\gamma l)}{Z_c + Z_T \tanh(\gamma l)}$$

which, in the case of lossless lines where $\gamma = j\beta$, again gives us expression (3.9).

It is possible, given Z_h, to deduce from it the characteristics of a line which should present the desired impedance $Z = Z(l)$. However, the solution is not direct. It is first necessary to find Z_c such that the moduli of the reflection coefficients associated with Z_h and Z are equal, and then to determine on a Smith chart the necessary rotation l/λ_m. Therefore, limiting ourselves to submultiples of the wavelength λ_m, we obtain the following expressions from relation (3.9):
for $l = \lambda_m/4$

$$Z\left(\frac{\lambda_m}{4}\right) = \frac{Z_c^2}{Z_T} \tag{3.10}$$

for $l = \lambda_m/8$ and $Z_T = 0$

$$Z\left(\frac{\lambda_m}{8}\right) = j\,Z_c \tag{3.11}$$

and for $l = \lambda_m/8$ and $Z_T = \infty$

$$Z\left(\frac{\lambda_m}{8}\right) = -jZ_c$$

or

$$\frac{1}{Z(\lambda_m/8)} = \frac{j}{Z_c} \tag{3.12}$$

Following these considerations concerning the application of microstrip technology, we are now in a position to calculate the characteristic impedance and the widths and lengths of the stubs required for the matching circuits.

Calculating the characteristic impedances

Input For $R_0 = 50\,\Omega$, which is the source impedance in its Thévenin representation, we wish to present an admittance $Y_s = G_s + jB_s$ equivalent to $\Gamma_s = 0.395\,(129°)$, or

$$Y_s = (25.62 - 18.63j)\ \text{mS}$$

and hence

$$\frac{1}{G_s} = 39\,\Omega \qquad \frac{1}{jB_s} = 53.7j\,\Omega$$

G_s is obtained by using a $\lambda_s/4$ stub with characteristic impedance Z_s such that (see Eqn (3.10))

$$Z_s = \left(\frac{R_0}{G_s}\right)^{\frac{1}{2}}$$

or

$$Z_s = 44.2\,\Omega$$

B_s is obtained by using a short-circuited $\lambda_s'/8$ stub with characteristic impedance (see Eqn (3.11))

$$Z_s' = 53.7\,\Omega$$

Output For a normalized load R_0 of $50\,\Omega$, it is necessary to present an admittance $Y_L = G_L + jB_L$ corresponding to $\Gamma_L = 0.747\,(+54°)$ or

$$Y_L = (3.63 - 9.92j)\ \text{mS}$$

and hence

$$\frac{1}{G_L} = 275\,\Omega \qquad \frac{1}{jB_L} = 100.8j\,\Omega$$

G_L is obtained by using a $\lambda_L/4$ stub with characteristic impedance $Z_L = (R_0/G_L)^{\frac{1}{2}}$ or

$$Z_L = 117.3\ \Omega$$

B_L is obtained by using a short-circuited $\lambda_L'/8$ stub with characteristic impedance

$$Z_L' = 100.8\,\Omega$$

Calculating the strip widths and lengths

We choose to fabricate these stubs on copper-clad double-sided board with Polyguide dielectrics of relative permittivity $\varepsilon_r = 2.32$ and height $h = 1.59$ mm (1/16 in). Referring to Fig. 3.21 we see that the $W_0//h$ ratio corresponding to a characteristic impedance $R_0 = 50\,\Omega$ is 3; hence

$$W_0 = 4.77\text{ mm}$$

Still using the graph to obtain the strip widths but employing Eqns (3.8a) and (3.8b) to calculate the wavelengths (to be precise, the wavelength λ_0 in air corresponding to the operating frequency of 2.5 GHz is 120 mm), we obtain the following results.

At the input, for $Z_s = 44.2\,\Omega$, $W_s/h = 3.65$ and $\lambda_0/\lambda_s = 1.4065$ or

$$W_s = 5.80\text{ mm} \qquad \lambda_s/4 = 21.33\text{ mm}$$

For $Z_s' = 53.7\,\Omega$, $W_s'/h = 2.65$ and $\lambda_0/\lambda_s' = 1.3928$ or

$$W_s' = 4.21\text{ mm} \qquad \lambda_s'/8 = 10.77\text{ mm}$$

At the output, for $Z_L = 117.3\,\Omega$, $W_L/h = 0.45$ and $\lambda_0/\lambda_L = 1.3317$ or

$$W_L = 0.72\text{ mm} \qquad \lambda_L/4 = 22.53\text{ mm}$$

For $Z_L' = 100.8\,\Omega$, $W_L'/h = 0.74$ and $\lambda_0/\lambda_L' = 1.3419$ or

$$W_L' = 1.18\text{ mm} \qquad \lambda_L'/8 = 11.18\text{ mm}$$

We now insert the results obtained on the diagram of Fig. 3.22 where the dimensions are in millimetres. We see that the resulting matching networks are not symmetrical about the 'propagation axes', although a circuit of this kind is

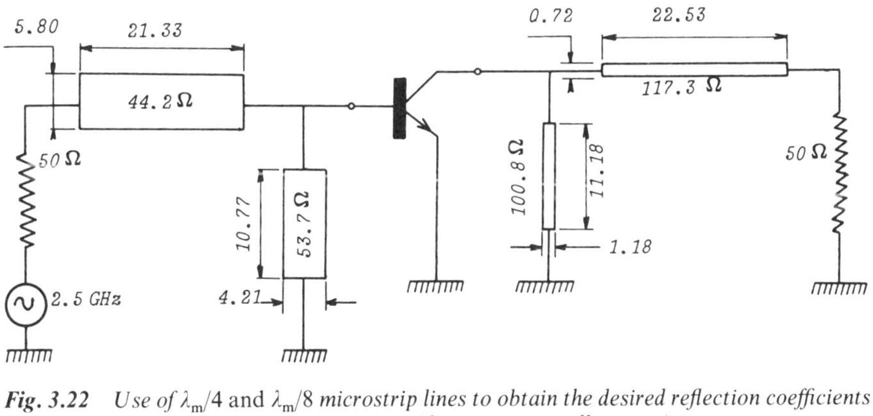

Fig. 3.22 *Use of $\lambda_m/4$ and $\lambda_m/8$ microstrip lines to obtain the desired reflection coefficients at the transistor ports (dimensions in millimetres).*

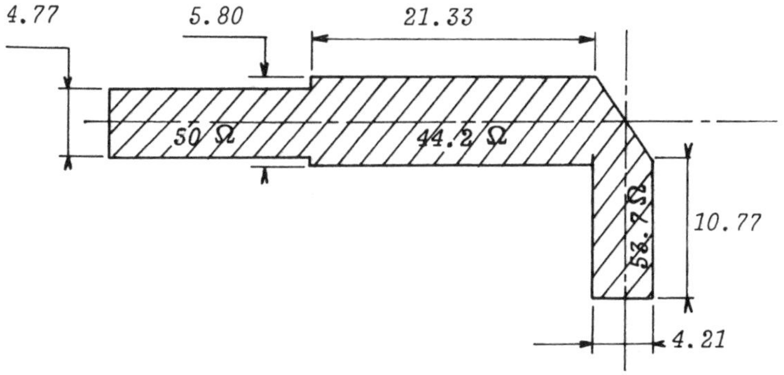

Fig. 3.23 Asymmetrical microstrip design at the input (dimensions in millimetres).

quite feasible. We give as an example the diagram of a Milar two-stub tuner for use at the input (Fig. 3.23).

A symmetrical structure is possible provided that stubs do not have an excessively high characteristic impedance, which is the case here at the input where $Z_s' = 53.7\,\Omega$. Thus, instead of presenting $53.7j\,\Omega$ in a single stub, we shall present $107.4j\,\Omega$ in two symmetrical stubs giving $W_s'' = 0.97$ mm and $\lambda_s''/8 = 11.24$ mm; this configuration is shown in Fig. 3.24.

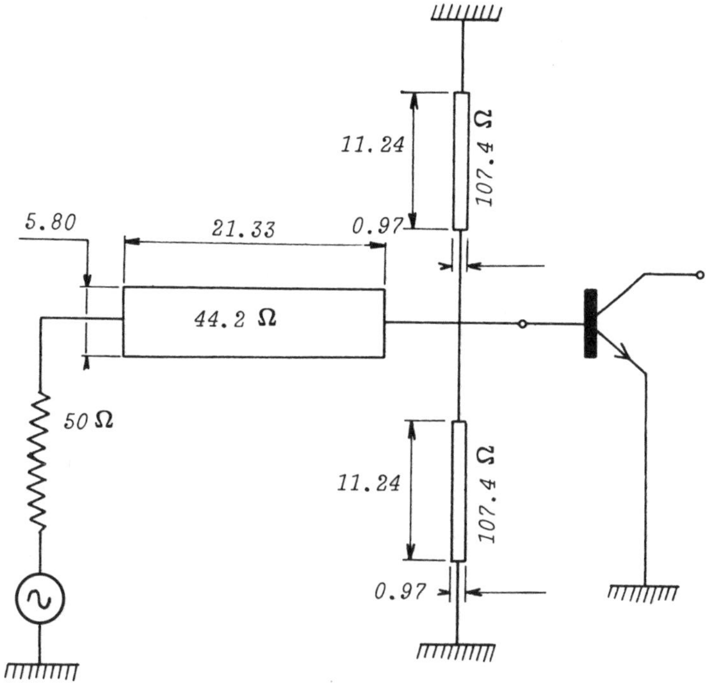

Fig. 3.24 Input circuit design based on the use of symmetrical stubs (dimensions in millimetres).

Having designed the matching circuits for a dielectric such that $\varepsilon_r = 2.32$ and $h = 1.59$ mm (1/16 in), it now remains to DC bias the transistor correctly, i.e. to prevent unwanted oscillations as well as mismatches and RF leakage. The last two problems can be overcome by applying the power at the end of a $\lambda_m/4$ line which is short-circuited by a capacitance; thus the other end presents an infinite impedance. An open-circuited $\lambda_m/2$ line supplied at its midpoint can also be used. Finally, the decoupling circuits can be constructed using, for example, 1 kpF capacitors. A circuit giving a power gain of better than 10 dB over a 500 MHz range centred on 2.5 GHz has been designed on this basis (Fig. 3.25).

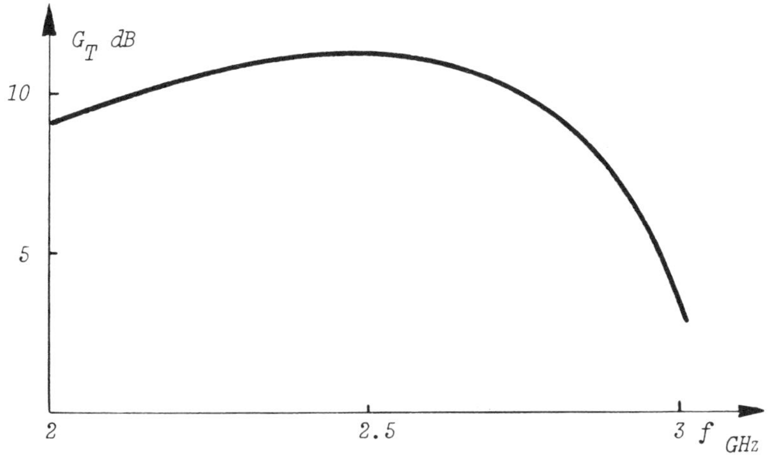

Fig. 3.25 *Power gain of an amplifier stage using distributed-element circuits.*

Better characteristics, in particular increased selectivity, could be obtained by using more expensive fibreglass-reinforced polytetrafluoroethylene-based substrates such as RT/Duroid 5880 or the equivalent P2M (OAK 700 for which $\varepsilon_r = 2.33$).

3.4. Maximum instability oscillators

THEORETICAL BASIS

A transistor of known scattering matrix S at the desired operating frequency is assumed to have a load impedance Z_L whose associated reflection coefficient is

$$\Gamma_L = \frac{Z_L - R_0}{Z_L + R_0}$$

The reflection coefficient S'_{11} seen at the input of the transistor with this load is therefore, from Eqn (2.10),

$$S'_{11} = S_{11} + \frac{S_{12}S_{21}\Gamma_L}{1 - S_{22}\Gamma_L}$$

as illustrated in Fig. 3.26. This is the case where an impedance $Z = R + jX$ is presented such that

$$S'_{11} = \frac{Z - R_0}{Z + R_0}$$

and the modulus is

$$|S'_{11}| = \frac{\{(R - R_0)^2 + X^2\}^{\frac{1}{2}}}{\{(R + R_0)^2 + X^2\}^{\frac{1}{2}}}$$

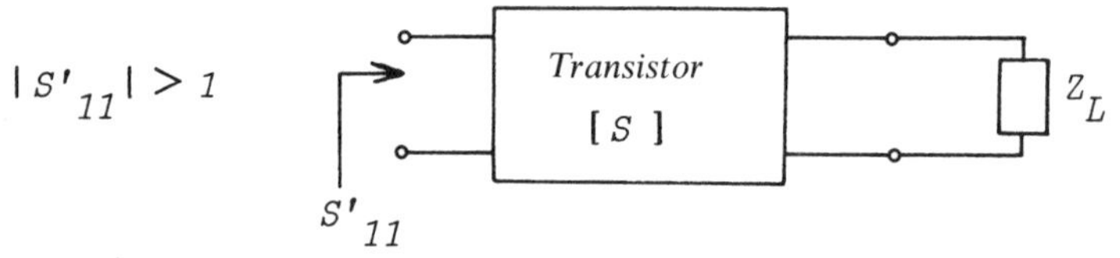

Fig. 3.26 Reflection coefficient presented to the oscillator input: oscillating condition.

The potential oscillation condition $R < 0$ can therefore be expressed in reflection coefficient terms by $|S'_{11}| > 1$; alternatively if the modulus of the reflection coefficient seen at the transistor input is less than unity, it will then be impossible to produce oscillation. This is exactly the case, irrespective of the Γ_L value, for an unconditionally stable transistor. In practice, we shall therefore select a transistor such that $|K| < 1$ (see discussion in Chapter 2, Section 2.2).

Having selected the transistor, the first step consists in plotting on a Smith chart the locus of the reflection coefficient Γ_L such that $|S'_{11}| = 1$ (we have seen in Chapter 2 that this is a circle) and determining the unstable region on the chart. We shall select Γ_L in this area.

In the second step, we deduce the value of the load impedance Z_s required at the input. For this purpose we can use the configuration shown in Fig. 3.27. When the source E is coupled we have the following relationships (see Eqn (2.2)) between

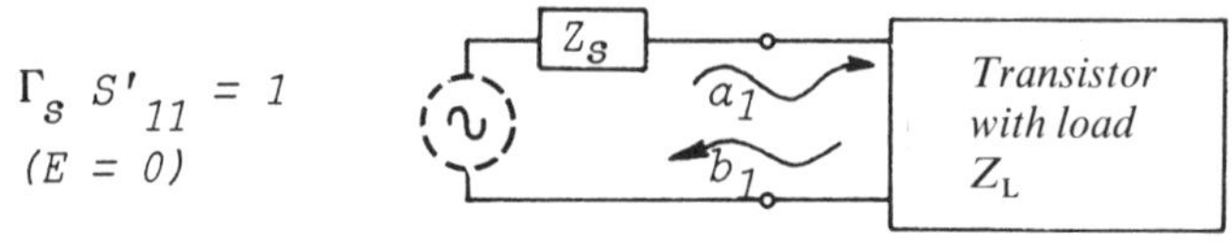

Fig. 3.27 Selection of the source impedance required to achieve tuned conditions.

the incident wave b_1 and the reflected wave a_1:

$$a_1 = \Gamma_s b_1 + \frac{R_0^{\frac{1}{2}}}{Z_s + R_0} E$$

where

$$\Gamma_s = \frac{Z_s - R_0}{Z_s + R_0}$$

The flowchart can be plotted directly and as shown in Fig.3.28. We obtain, for example by using Mason's rule (Appendix A),

$$\frac{b_1}{b_s} = \frac{S'_{11}}{1 - \Gamma_s S'_{11}}$$

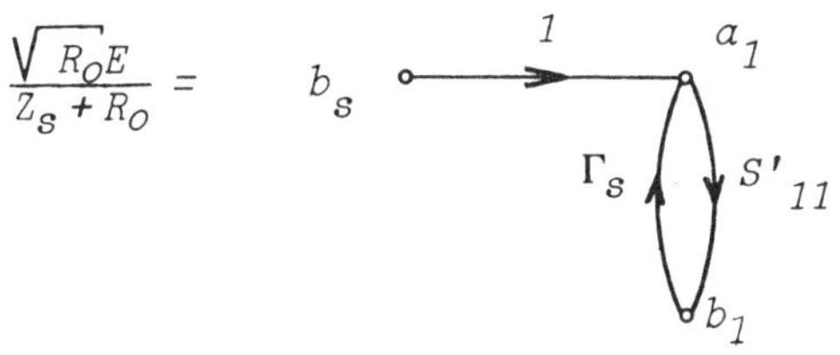

Fig. 3.28 *Graph obtained from Figs 3.26 and 3.27.*

(This is Eqn (A3)). Oscillation will therefore occur, i.e. there is a reflected wave b_1 even when $E = 0$, for a reflection coefficient at the required frequency such that $\Gamma_s S'_{11} = 1$ or

$$\Gamma_s = \frac{1}{S'_{11}} \tag{3.13}$$

Having determined Γ_L and Γ_s, we are now in a position to design an oscillator.

Typical investigation with a naturally unstable transistor: finding the maximum instability

It is in fact difficult to find a transistor with $|K| < 1$, as is shown by the three examples in Table 3.5. Even in the common base configuration, these transistors remain unconditionally stable. However, there are certain devices, e.g. the HXTR

Table 3.5

Transistor	Frequency (GHz)	K(common emitter)
BFT 65	1.8	1.197
BFR 90	1.8	1.136
BFR 34 A	1.8	1.125

4101 from Hewlett packard, which are specially designed for use as oscillators. In the common base configuration the scattering parameters at 2 GHz are

$$S_{11} = 0.964\,(144°)$$
$$S_{21} = 1.95\,(-59°)$$
$$S_{12} = 0.039\,(120°)$$
$$S_{22} = 1.068\,(-45°)$$

We obtain

$$\Delta = 0.97075\,(101.8°)$$

Hence

$$|\Delta|^2 = 0.94236$$

and therefore

$$K = -0.83865$$

We thus indeed have $|K| < 1$, and so we proceed to the first step which consists of drawing the input instability circle defined by

$$R_2 = \frac{|S_{12}S_{21}|}{||S_{22}|^2 - |\Delta|^2|}$$

$$\Omega_2 = \frac{(S_{22} - \Delta S_{11}{}^*)^*}{|S_{22}|^2 - |\Delta|^2}$$

Numerically this gives

$$R_2 = 0.38358 \qquad \Omega_2 = 0.71084\,(63.9°)$$

We can show that $|\Omega_2| - R_2$ is greater than zero and thus that the circle does not contain the centre of the Smith chart, which was to be expected given that $|\Delta| < |S_{22}|$. Moreover, since $|S_{11}|$ is less than unity, the area inside the circle is an unstable region, and it is here, provided that we stay within the chart, that we must select Γ_L. The question is now to decide where.

It seems natural, assuming *a priori* the uniqueness of the solution, to place ourselves at a point of maximum instability, i.e. such that $|S'_{11}|$ is as large as possible, provided that this point still lies within the chart. In more general terms, this means that we have to find the locus of the reflection coefficients Γ_L presenting a coefficient S'_{11} of modulus k greater than unity at the input. The calculations given in Appendix C yield the following result: the locus of Γ_L such that $|S'_{11}| = k$ is a circle

$$\text{radius } R(k) = \frac{k|S_{12}||S_{21}|}{|k^2|S_{22}|^2 - |\Delta|^2|}$$

$$\text{centre } \Omega(k) = \frac{(k^2 S_{22} - \Delta S_{11}{}^*)^*}{k^2|S_{22}|^2 - |\Delta|^2} \tag{3.14}$$

We have shown, moreover, that angle $\widehat{OPK}$, where O is the centre of the Smith chart, P is the point representing $\Omega(1)$ and K is the point representing $\Omega(k)-k>1$, remains constant and equal to θ_1 as k varies. Thus, the centres are aligned and θ_1 is obtained from the relation

$$2|\Omega(1)|\cos\theta_1 = \frac{(|S_{22}|^2-|\Delta|^2)(|\Delta|^2-|S_{11}|^2|S_{22}|^2)+(|S_{22}|^2+|\Delta|^2)|S_{12}|^2|S_{21}|^2}{(|S_{22}|^2-|\Delta|^2)|S_{12}||S_{21}||\Delta||S_{22}|} \tag{3.15}$$

Table 3.6

k	$R(k)$	$\Omega(k)$
1	0.3836	0.7108 (63.9°)
1.05	0.2534	0.7838 (55.7°)
1.10	0.1911	0.8235 (52.3°)
1.20	0.1303	0.8641 (49.3°)
1.35	0.0903	0.8912 (47.6°)
1.5	0.0702	0.9045 (46.8°)
2	0.0420	0.9220 (45.8°)
3	0.0245	0.9307 (45.3°)
∞	0.0	0.9363 (45.0°)

The numerical representation of the transistor in question is given in Table 3.6. $R(k)$ decreases rapidly as a function of k ($dR/dk<0$) and tends to zero as k tends to infinity (Fig. 3.29). The centres $\Omega(k)$ are plotted on a Smith chart in Fig. 3.30. They effectively move along a straight line. The calculation using Eqn (3.15) gives $\theta_1 = 120.47°$ as we can show. We note the important result $\Omega(\infty) = 1/S_{22}$, and as $R(\infty) = 0$ we have solved our problem. The point maximizing the input instability is that rendering it infinite; the point $\Gamma_L(\infty)$ is such that

$$\Gamma_L(\infty) = \frac{1}{S_{22}} \tag{3.16a}$$

This solution is of course unique; it was obtained merely by examining $|S'_{11}|$. It leads to (the second step)

$$\Gamma_s(\infty) = 0 \tag{3.16b}$$

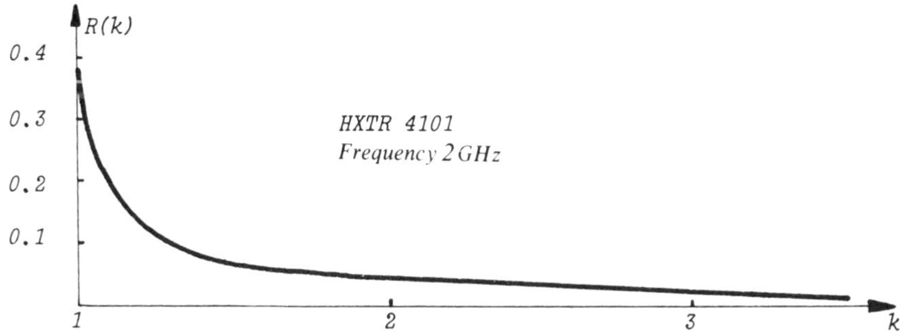

Fig. 3.29 *Decrease in the radius of the instability circles as a function of the coefficient k.*

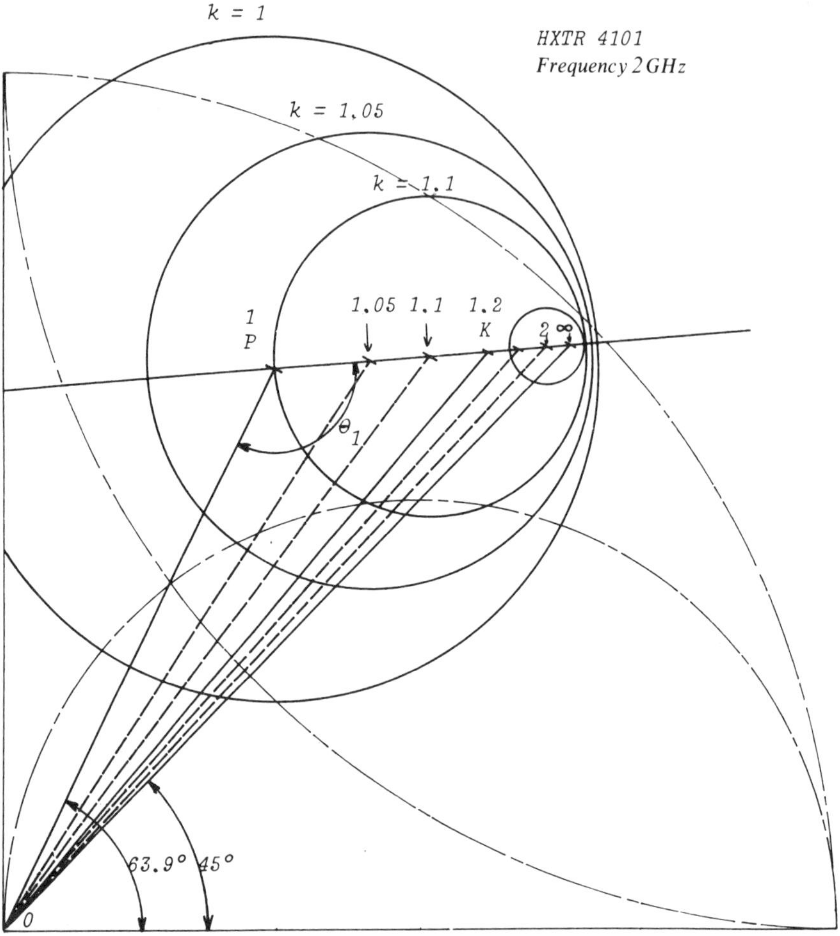

Fig. 3.30 Locus of the reflection coefficients Γ_L of a load presenting a given instability k at the input.

It is therefore unnecessary to provide an input matching circuit. However, let us remember that these solutions are only valid because $|S_{22}| > 1$; if the reverse were true, we would obtain the configuration shown in Fig. 3.31.

The maximum instability value which we can therefore obtain is that for which the circle $|S'_{11}|$ = constant is externally tangential to the Smith chart. Let k_t be this value. We then have the relation

$$|\Omega(k_t)| = R(k_t) + 1$$

giving the value of k_t. The calculations given in Appendix C yield the expression

$$k_t = \frac{|S_{12}||S_{21}| + \{|S_{12}|^2|S_{21}|^2 - (1-|S_{22}|^2)(|\Delta|^2 - |S_{11}|^2)\}^{\frac{1}{2}}}{1-|S_{22}|^2} \tag{3.17}$$

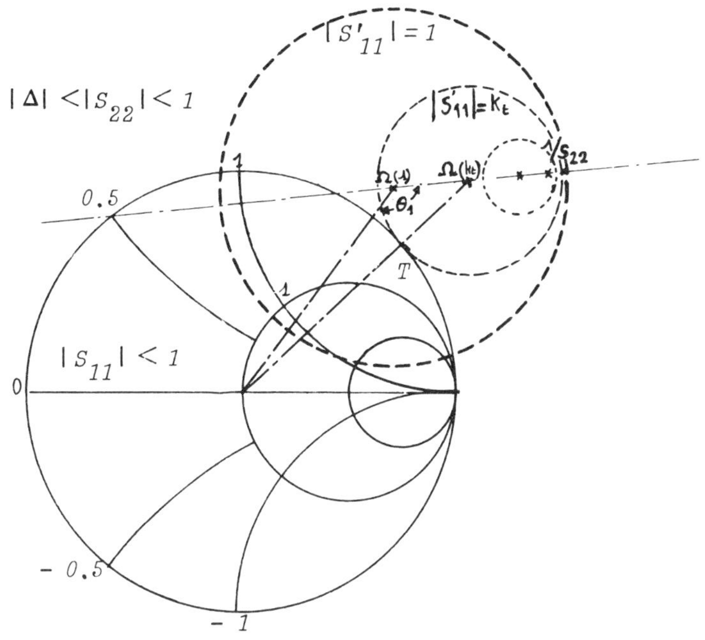

Fig. 3.31 *Instability circles for the case where S_{22} is less than unity (tangential instability circle).*

Thus, in the case where $|S_{22}|$ is less than unity, the reflection coefficient Γ_L which must be adopted is

$$\Gamma_L(k_t) = 1\,(\text{angle } \Omega(k_t)) \tag{3.18}$$

which is point T on Fig. 3.31.

Note that these considerations derived from analysis of an oscillator based on the HXTR 4101 device are for the configuration $|\Delta| < |S_{22}|$.

DESIGN USING A TRANSISTOR FOR WHICH $|\Delta| > |S_{22}|$

This involves the LAE 2001R (the former designation was RTC 101 L). In the band of frequencies centred on 1.9 GHz that are of interest to us, this transistor is unconditionally stable in both the common base and the common emitter configurations. However, inserting a low value capacitance (1 pF) between the collector and the base of the common emitter circuit enables us to create a zone of unstable operation artificially. In fact, moving from the experimental plane to the model plane (Eqn (1.58)), we have obtained the following values at 1.9 GHz:

$$S_{11} = 0.47315\,(-37^\circ)$$
$$S_{22} = 0.1122\,(+87^\circ)$$
$$S_{21} = 1.2589\,(-117^\circ)$$
$$S_{12} = 0.5309\,(-74^\circ)$$

Hence

$$\Delta = 0.6956\,(-7.2°)$$

and we therefore obtain

$$K = 0.9332 < 1 \qquad |\Delta| > |S_{22}|$$

The circles parametrized as values of k are shown in Fig. 3.32. Note the critical instability ($k = 1$) and tangential instability ($k = k_t$) circles.

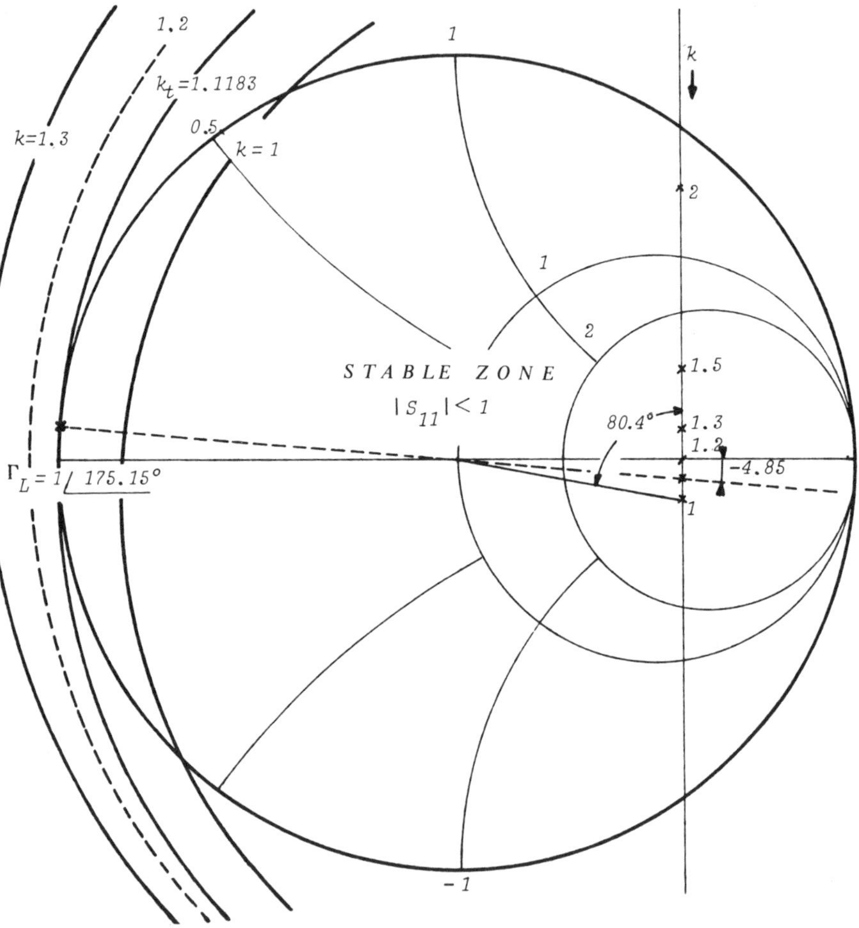

Fig. 3.32 Load reflection coefficient Γ_L obtained from the tangential instability circle in the case where $|\Delta| > |S_{22}|$.

For the special value $k_t = 1.1183$, circle C_{k_t} is tangential to the Smith chart. We compute

$$R(k_t) = 1.5965 \qquad \Omega(k_t) = 0.5965\,(-4.85°)$$

These three values follow from Eqns (3.17) and (3.14). In fact the relation giving k_t remains valid for $|\Delta| > |S_{22}|$ (see Appendix C, Section C3). The value of the reflection coefficient which the load Z_L should present is therefore

$$\Gamma_L(k_t) = 1\,(175.15°)$$

Numerical substitution then gives $S'_{11} = 1.1182\,(-28.35°)$, whence we derive

$$\Gamma_s(k_t) = 0.8943\,(28.35°)$$

The problem is now to determine whether impedance Z_L corresponding to $\Gamma_L(k_t)$ or admittance Y_s corresponding to $\Gamma_s(k_t)$ can be realized in microstrip technology, i.e. whether it may be necessary to have excessively large or small strip widths. Now, the impedance Z_L is equivalent to 2.1j Ω. In order to present this value it will be necessary to have a short-circuited $\lambda_m/8$ line with characteristic impedance $Z_c = 2\,\Omega$, i.e. of prohibitive width. In practice, we can avoid this by simply providing a short circuit (e.g. a 100 pF capacitor) on the output side of the prototype. Moreover, either by using the relation

$$Y = \frac{1}{R_0}\,\frac{1-\Gamma}{1+\Gamma}$$

or, which is less accurate, by using the Smith chart, we find

$$Y_s = (1.188 - 5.035\mathrm{j})\,\mathrm{mS}$$

Thus, as we saw in Section 3.3, with $R_0 = 50\,\Omega$ we shall initially present $1/G_s$ using a $\lambda_m/4$ stub of characteristic impedance $Z_s' = (R_0/G_s)^{\frac{1}{2}} = 205\,\Omega$; then a short-circuited $\lambda_m/8$ stub of impedance $Z_s'' = -1/B_s = 198\,\Omega$ connected in parallel will present the imaginary part $1/\mathrm{j}B_s$.

Provided that we use a substrate of sufficiently low relative permittivity, characteristic impedances of the order of 200 Ω are still easily realized. For $\varepsilon_r = 2.3$ and a substrate height of 1.59 mm, we obtain a width of approximately 0.16 mm. In the case when the substrate is specified and when, unfortunately, ε_r is excessively high, it will be necessary to revise the choice of the pair $\{\Gamma_L; \Gamma_s\}$. Note that a slight rotation of the phase Γ_L has comparatively little effect on the instability. For $\Gamma_L = 1\,(+150°)$, we obtain $\Gamma_s = 0.9114\,(+42°)$. Thus $Z_L = 13\mathrm{j}\,\Omega$ and in this case it is advisable to use two stubs of impedance $Z_c = 26\,\Omega$ connected in parallel. Then

$$Y_s = (1.14 - 7.8\mathrm{j})\,\mathrm{mS}$$

and the real part will require a stub with $Z_s' = 209\,\Omega$ which is difficult to implement. The achievement of $Z_s'' = 128\,\Omega$ for the imaginary part poses no problem.

An economic design using these values has been found to be possible with epoxy of relative permittivity 4.9 and thickness 1.59 mm (1/16 in). The 209 Ω stub consisted of a copper conductor mounted in air 12 mm from the base of the case, which was also made of copper. The dimensions (widths and lengths of the $\lambda_m/4$ and $\lambda_m/8$ stubs) computed on the basis of these data are shown in Fig. 3.33. The supply voltage was stabilized at 15 V and we measured

$$V_{CE} = 9.45\,\mathrm{V} \qquad V_{BE} = 0.75\,\mathrm{V} \qquad I_c = 14.2\,\mathrm{mA}$$

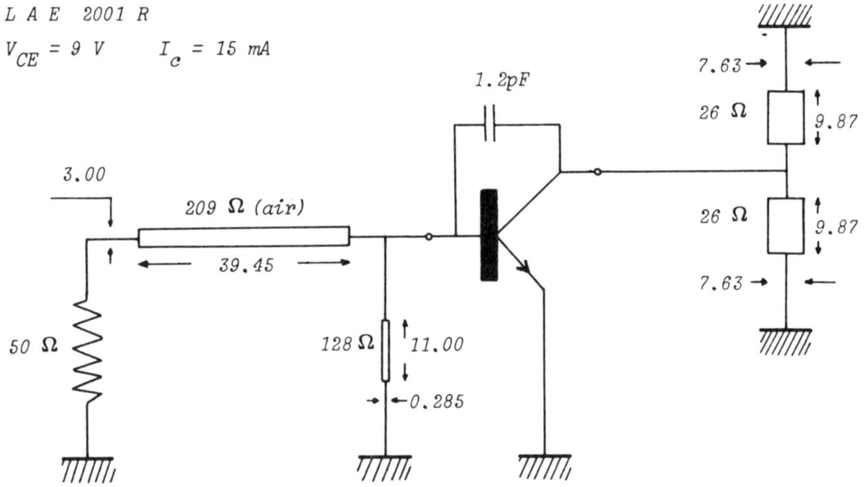

Fig. 3.33 *1.9 GHz oscillator: microstrip realization of the reflection coefficients at the ports (dimensions in millimetres).*

We have obtained, without a practical experiment, the desired oscillation frequency. The curves given in Fig. 3.34 show the changes in this frequency and in the power delivered as a function of the supply voltage.

In conclusion, at 1.9 GHz the oscillator delivers approximately 10 mW into a 50 Ω resistive load. Using a 15 V supply we measure a 'pushing' of 3.5 MHz V^{-1} (an alternative design enabled us to go down to less than 0.5 MHz V^{-1}). However, when the voltage standing wave ratio increases from 1 to 1.5 (75 Ω load) a 'pulling' of 1.2 MHz is obtained.

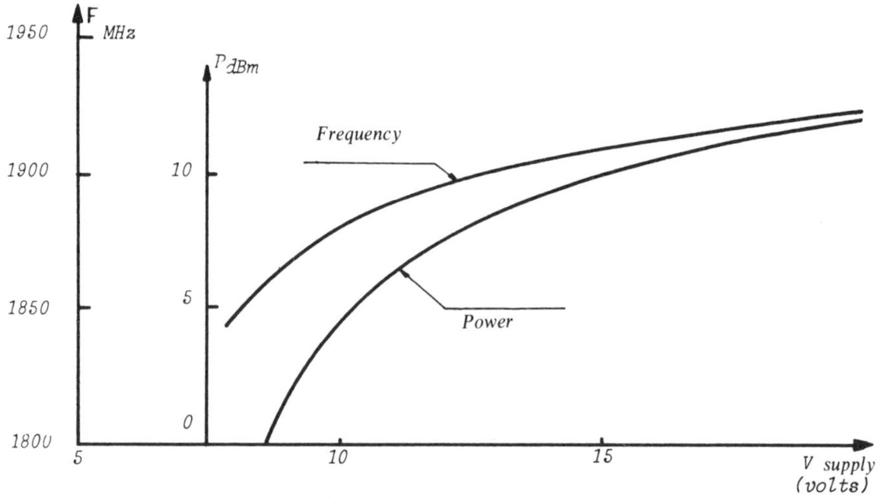

Fig. 3.34 *Effect of the supply voltage on the frequency and power delivered by the oscillator.*

Chapter 4

S parameter and noise measurement

4.1. Network analyser and the automatic measurement of scattering parameters

PRINCIPLE OF THE NETWORK ANALYSER

A network analyser enables us to evaluate the reflection coefficients S_{11} and S_{22} and the transmission coefficients S_{12} and S_{21} of a two-port network as a function of frequency. More precisely, it produces the amplitude and phase of these four complex parameters. The measurements performed are essentially relative in nature because they are based on a comparison of the reflected or transmitted wave with the incident wave at the port selected. Determination of the phase will therefore require precise location of the reference planes validating the measurement; these can be used, if necessary, for performing the matrix correction described in Eqn (1.58). We immediately have at our disposal a simple calibration procedure for reflection measurements for example. For the reference plane a load $R_0 = 50\,\Omega$, i.e. a purely resistive load, must give a zero reflection coefficient corresponding to the centre of the Smith chart, a short circuit must effectively be represented by point 1 (180°) and an open circuit will be represented by the point 1 (0°) diametrically opposite.

SYSTEM COMPONENTS

The signal source

The signal source can be a sweep generator or a synthesizer. The generator is necessary for characterizing wide band devices. The synthesizer, which is capable of supplying a very pure and highly stable signal, is ideal for performing measurements on narrow band filters, for example, and generally for any one- or two-port network whose amplitude and phase vary rapidly as a function of frequency.

The signal separation unit

By virtue of the principle of the equipment, the signal delivered by the source must follow two different electrical paths. A divider bridge, whose essential feature is its wide passband, can be used for this purpose. It is followed, in the path which we shall call the **reference path**, by an adjustable attenuator and stub whose roles will be described later. The incident signal follows the other path.

The first directional coupler

The incident signal is applied to the port selected as the input, whereas the device under test is connected to the free end of the main transmission line. A fraction of the reflected signal will therefore be present at the measuring port of the coupled line. This fraction represents the coupling which is compensated by the presence of the attenuator in the reference path.

The second directional coupler

At the outset there is nothing to prevent the signal transmitted by the two-port network under test from being measured directly after it has undergone an attenuation of the same value as that of the coupling envisaged for the reflection test, if we wish to be able to compare it with the reference signal. Thus it appears advisable to load the output of the two-port network with a second coupler identical with the first and to take the transmitted signal from the coupled path.

The two switches

The circuit described above equipped with a switch C_2 enables S_{11} and S_{21} measurements to be performed without further manipulation. Addition of a second switch C_1 enables the other two parameters to be determined without the need to turn the two-port network under test.

At this point in our description we arrive at the functional unit shown in Fig. 4.1 which, in a restricted sense, is known as a network analyser (see below). Note the presence of the adjustable stub which we have already mentioned. In fact, an air line of this kind of length ΔL has to compensate for the difference between the electrical routes along the test path and the reference path which causes a phase shift $\Delta\varphi$ proportional to the frequency. The compensation will effectively be achieved if we satisfy the relation

$$\Delta\phi = \left(\frac{2\pi}{c}\Delta L\right)f \tag{4.1}$$

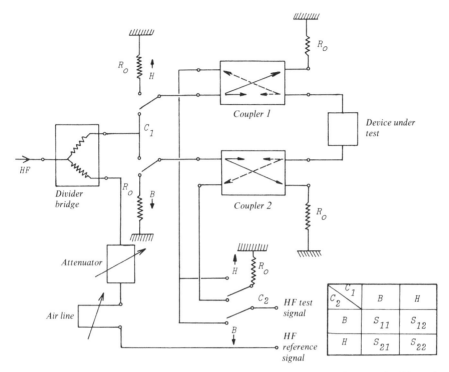

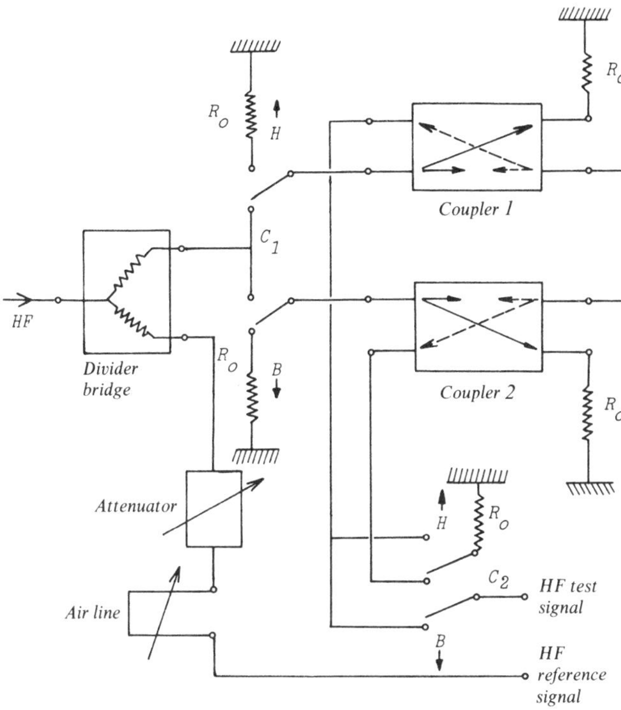

C_2 \ C_1	B	H
B	S_{11}	S_{12}
H	S_{21}	S_{22}

Fig. 4.1 *System for automatically selecting high frequency test signals. A truth table is also included.*

where c is the propagation velocity in the air line; if the line were filled with dielectric of relative permittivity ε_r, it would be necessary to multiply the length by $\varepsilon_r^{\frac{1}{2}}$. In the case of automatic measurement, however, the electrical length of the test path and its attenuation should remain constant whatever parameter is selected. For example, the transmission calibration, which is accomplished by connecting the two measuring ports directly, should theoretically give the point 1 (0°) irrespective of frequency.

Let us return to our system. In order for it to operate it must also incorporate a receiver and a visual display unit.

The receiver

When the analyser is operating at low frequency (e.g. up to a maximum of 100 kHz), the test and reference signals (which may, in contrast, reach several tens of gigahertz (40 GHz for commercial equipment)) must be converted into an **intermediate frequency** (IF) signal before any processing is carried out. For this purpose, it is necessary to generate a local oscillator at either the sum or the difference frequency of the input signal and the IF. This local oscillator is applied to mixers in the test and reference paths. The principle of a system initially using two fixed local oscillators accurately offset from the IF is illustrated in Fig. 4.2. These two oscillators each drive a mixer whose other input is loaded by the microwave source. Receivers operating at a few gigahertz can be based on this principle of mixing with the fundamental frequency which undoubtedly ensures optimum performance in terms of both sensitivity and dynamic range. For analysers operating at several gigahertz the test and reference signals, whose frequency is that of the sweep generator, can be mixed with the harmonics of the local oscillator which, by means of a phase-locked loop, is able to track the source. The technical problem is then to be able to achieve continuous tracking, apart from the IF offset, even when the harmonic is changed in order to have sweep widths which are not limited to one octave.

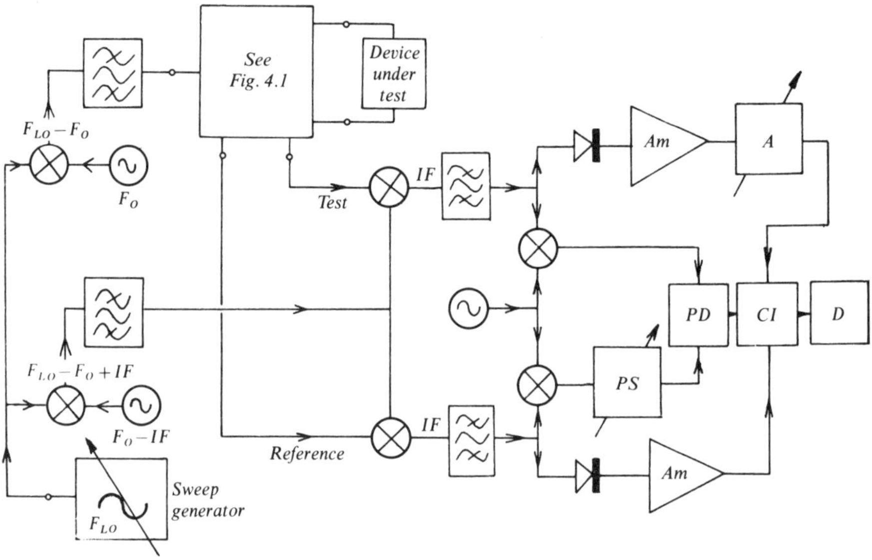

Fig. 4.2 *Network analyser: example of a high sensitivity receiver: A, attenuator; Am, amplifier; PS, phase shifter; PD, phase detector; CI, computer interface; D, display.*

In addition to the frequency change, which provides IF signals carrying the same relative amplitude and phase difference information as the test and reference paths, the receiver must also perform detection and comparison functions leading to the measurement of the modulus and argument of the scattering parameters. Therefore detectors mounted at the output of the IF filters are followed by logarithmic or linear amplifiers which, after taking the difference or ratio, will provide the modulus expressed in decibels in the former case.

The presence, in the test path, of a calibrated attenuator which is used in the situation when the device inserted has a high gain should be noted. In order to determine the phase a second frequency conversion is generally performed, which makes it possible to go down from a few tens of megahertz to a few hundred kilohertz. A comparator then restores the phase shift between the test and reference signals from -180° to $+180^\circ$. This configuration is shown in Fig. 4.2. The presence of a calibrated electronic phase shifter should be noted.

We can also use synchronous detection which enables the modulus and argument to be measured simultaneously. Two balanced mixers are used for this purpose. The first receives the reference signal $R\cos(\omega t)$ and the test signal $T \times \cos(\omega t + \theta)$; we therefore obtain, after low pass filtering, a voltage proportional to $T\cos\theta$. The same reference signal $R\cos(\omega t)$ will always be present at the second mixer, whereas the test signal will undergo a 90° phase shift before being applied, this time giving a voltage proportional to $T\sin\theta$. Synchronous detection is therefore ideally suited to a polar representation.

Display

One possibility for the display is polar representation, whose main drawbacks are the low display dynamics and the absence of frequency information; this can be obtained, however, by using markers. Alternatively the display of the Smith chart on the screen makes it easy to read off the normalized impedance from the reflection coefficient. A cartesian representation gives the variations in phase or amplitude as a function of frequency. We can use either a logarithmic or a linear scale in these last two quantities.

COMPENSATION FOR ERRORS INHERENT IN THE SYSTEM IN ORDER TO FIND THE TRUE VALUES

The directional coupler is at the heart of the network analyser described above. It dictates the extreme operating frequencies of the equipment and is of the exponential type if a very wide band is required. Furthermore, it will set the maximum directivity of the device, which determines the error made in measuring the S parameters. Thus, before discussing the various sources of error affecting this measurement, we shall examine the picture of the true reflection coefficient given by the non-ideal coupler.

The scattering matrix of a real coupler is expressed by

$$[S] = \begin{bmatrix} S_{11} & S_{12} & S_{13} & S_{14} \\ S_{12} & S_{11} & S_{14} & S_{13} \\ S_{13} & S_{14} & S_{11} & S_{12} \\ S_{14} & S_{13} & S_{12} & S_{11} \end{bmatrix}$$

where the reciprocity is taken into account and a double formal symmetry is assumed (Fig. 4.3).

Each port presents a reflection coefficient which must satisfy $|S_{11}| \ll 1$ when the other three ports are terminated by the reference impedances which in practice are identical and equal to R_0. The existence of an isolated path necessarily implies $|S_{12}| \ll 1$. When the ports are correctly terminated an incident wave a_1 is transmitted at port 1. At port 3, which is the main output, we obtain b_3

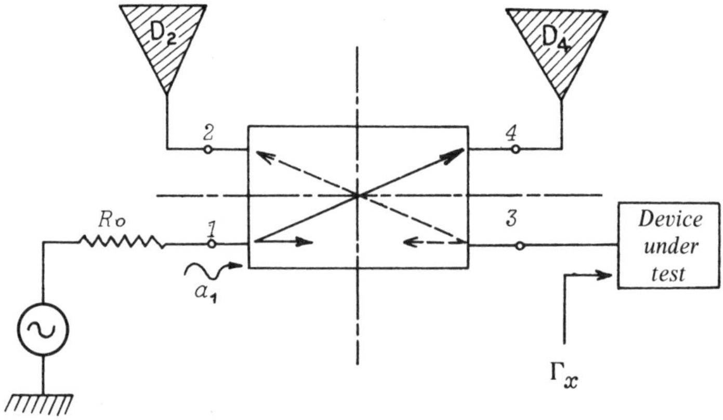

Fig. 4.3 Reflection coefficient measurement using a directional coupler.

$= S_{13}a_1$; the transmission T expressed in decibels is defined by

$$T_{\mathrm{dB}} = 20 \log |S_{13}| \tag{4.2}$$

At port 4, which is the coupled output, we obtain $b_4 = S_{14}a_1$; the coupling C expressed in decibels is defined by

$$C_{\mathrm{dB}} = 20 \log |S_{14}| \tag{4.3}$$

However, at port 2, which is the isolated output, a wave b_2 appears which, in terms of b_4, is given by

$$b_2 = \frac{S_{12}}{S_{14}} b_4$$

The directivity D is therefore

$$D_{\mathrm{dB}} = 20 \log \left| \frac{S_{14}}{S_{12}} \right| \tag{4.4}$$

We now connect detectors of impedance R_0 at ports 2 and 4 and a load of unknown reflection coefficient Γ_x at port 3. This circuit is shown in Fig. 4.3 where

$$a_2 = a_4 = 0 \qquad a_3 = \Gamma_x b_3$$

The matrix equation $[b] = [S][a]$ leads to the equalities

$$\begin{aligned}
b_1 &= S_{11}a_1 + S_{13}\Gamma_x b_3 && \text{(a)}\\
b_2 &= S_{12}a_1 + S_{14}\Gamma_x b_3 && \text{(b)}\\
b_3 &= S_{13}a_1 + S_{11}\Gamma_x b_3 && \text{(c)}\\
b_4 &= S_{14}a_1 + S_{12}\Gamma_x b_3 && \text{(d)}
\end{aligned}$$

From (c) we obtain

$$b_3 = \frac{S_{13}}{1 - S_{11}\Gamma_x} a_1$$

Hence in (b)

$$b_2 = \left(S_{12} + \frac{S_{13}S_{14}\Gamma_x}{1 - S_{11}\Gamma_x}\right)a_1$$

and in (d)

$$b_4 = \left(S_{14} + \frac{S_{12}S_{13}\Gamma_x}{1 - S_{11}\Gamma_x}\right)a_1$$

On taking the ratio and introducing the directivity D, we obtain

$$\frac{b_2}{b_4} = \frac{1}{D}\frac{1 + (S_{13}D - S_{11})\Gamma_x}{1 + (S_{13}/D - S_{11})\Gamma_x} \tag{4.5a}$$

This relation theoretically enables us to find Γ_x, and if the coupler had been ideal we would quite simply have obtained

$$\frac{b_2}{b_4} = S_{13}\Gamma_x \tag{4.5b}$$

Relation (4.5a) can be expanded as follows:

$$\frac{b_2}{b_4} = \frac{1}{D}\{1 + (S_{13}D - S_{11})\Gamma_x\}\left\{1 - \left(\frac{S_{13}}{D} - S_{11}\right)\Gamma_x + \left(\frac{S_{13}}{D} - S_{11}\right)^2 \Gamma_x{}^2 - \dots\right\}$$

For a good coupler the modulus of $S_{13}/D - S_{11}$ is small compared with unity; restricting ourselves to a second-order expansion, we shall write

$$\frac{b_2}{b_4} = \frac{1}{D} + S_{13}\left(1 - \frac{1}{D^2}\right)\Gamma_x - \left\{\frac{S_{13}{}^2}{D}\left(1 - \frac{1}{D^2}\right) - S_{13}S_{11}\left(1 - \frac{2}{D^2}\right)\right\}\Gamma_x{}^2 \tag{4.6}$$

Because of (4.5b), we can easily obtain the error term $\Delta(S_{13}\Gamma_x)$ whose modulus satisfies the inequality

$$|\Delta(S_{13}\Gamma_x)| \leqslant \frac{1}{|D|}(1 + |S_{13}{}^2||\Gamma_x{}^2|) + |S_{13}||S_{11}||\Gamma_x{}^2| \tag{4.7}$$

The error therefore appears as the sum of a first term due to the finite directivity of the coupler and a second term due to its own reflection. We note that for small reflection coefficients the error will in practice be equal to the reciprocal of the directivity.

In practice, we tend to consider the system from outside and to represent it by a flowchart taking into account all the possible paths which the incident wave may follow before producing the reflected or transmitted wave. We therefore find 2×6 error terms depending on whether switch C_1 of Fig. 4.1 is 'down', allowing the forward wave to pass into the two-port network under test, or 'up', which corresponds to the reverse wave. These terms are as follows.

The directivity error E_{DF} or E_{DR}

This error arises because part of the incident signal is present in the path assigned to measuring the reflected signal. This can be attributed to the non-infinite directivity of the coupler; however, this explanation disregards various other leaks, particularly the inevitable reflection due to the connection to the device under test.

Source mismatch error E_{SF} or E_{SR}

The source presents a reflection coefficient because its impedance is not strictly equal to the reference impedance.

Load mismatch error E_{LF} or E_{LR}

In the automatic analysers of the type shown in Fig. 4.1, the second port of the unit under test is terminated by the equipment itself which may also present a slight reflection coefficient.

Morphology error E_{RF} or E_{RR}

The reference and test paths are made up of different components which cause signal level variations as a function of frequency.

Transmission error E_{TF} or E_{TR}

This indicates the fluctation, as a function of frequency, in the signal level transmitted.

Radiation error E_{XF} or E_{XR}

Even if the analyser output is disconnected from the item under test, some transmitted energy will still be detected.

An error model of this type leads to the flowcharts shown in Figs 4.4(a) and 4.4(b) in which the S parameters are those of an unknown two-port network. This model is only useful and only provides the true S_{11}, S_{12}, S_{21} and S_{22} parameters if we are able to measure each of the six error terms associated with the direct and reverse configurations. We shall only describe how to measure the terms with second subscript F in Fig. 4.4(a); given symmetry, the principle for measuring the terms whose second subscript is R (Fig. 4.4(b)) is identical. We initially determine the terms E_{DF}, E_{SF} and E_{RF} by means of three manual reflection measurements (switches C_1 and C_2 of Fig. 4.1 are 'down'). For a one-port network load of coefficient Γ we eventually obtain the simplified flowchart of Fig. 4.5, since $S_{12} = S_{21} = 0$.

The measured coefficient Γ_m is related to the true coefficient Γ by

$$\Gamma_m = E_{DF} + \frac{E_{RF}}{1/\Gamma - E_{SF}} \tag{4.8a}$$

and, conversely,

$$\Gamma = \frac{\Gamma_m - E_{DF}}{E_{SF}(\Gamma_m - E_{DF}) + E_{RF}} \tag{4.8b}$$

As the one-port network is, in turn, a matched load ($\Gamma = 0$), a short circuit

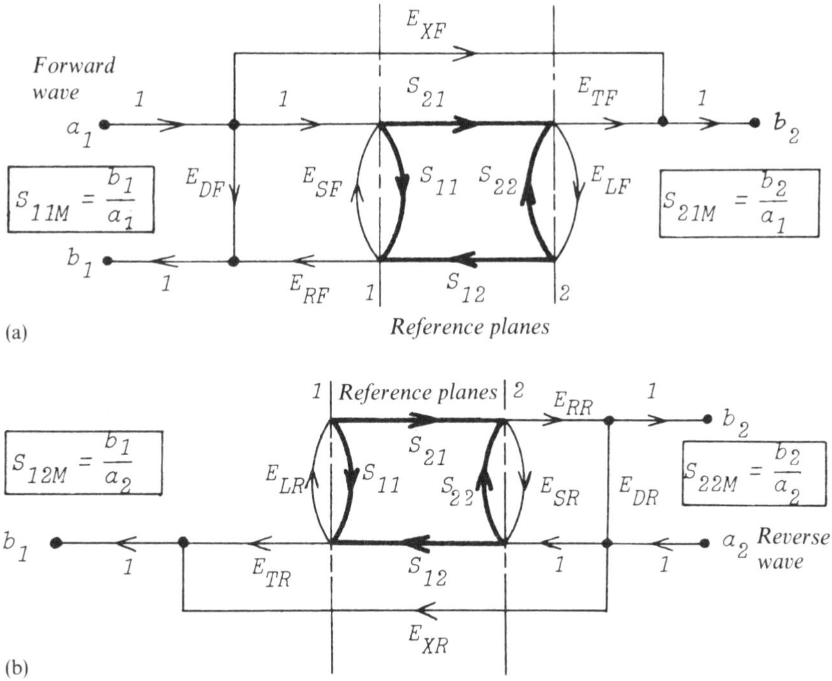

Fig. 4.4 *(a) Flowchart representing the path of the forward wave for automatic measurement of the S_{11M} and S_{21M} parameters; (b) flowchart representing the path of the reverse wave for automatic measurement of the S_{22M} and S_{12M} parameters.*

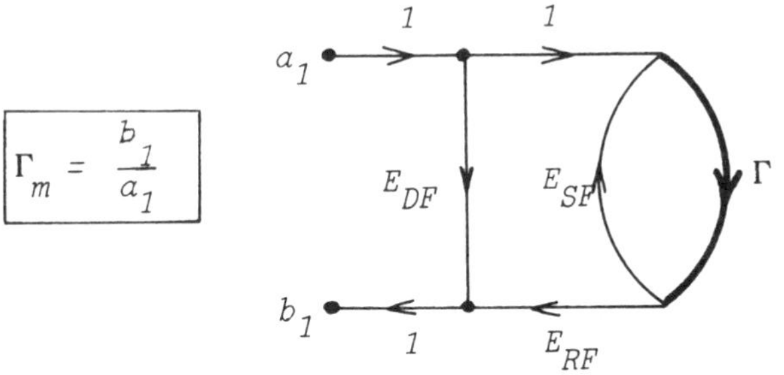

Fig. 4.5 *Obtaining the error terms E_{DF}, E_{RF} and E_{SF} by using manual reflection measurements.*

($\Gamma = -1$) and an open circuit ($\Gamma = 1$), we measure $\Gamma_m(0)$, $\Gamma_m(-1)$ and $\Gamma_m(1)$ respectively. Using (4.8a) we then obtain

$$E_{DF} = \Gamma_m(0) \tag{4.9a}$$

$$E_{SF} = \frac{\Gamma_m(1) + \Gamma_m(-1) - 2\Gamma_m(0)}{\Gamma_m(1) - \Gamma_m(-1)} \tag{4.9b}$$

$$E_{\mathrm{RF}} = -2\frac{\{\Gamma_{\mathrm{m}}(1)-\Gamma_{\mathrm{m}}(0)\}\{\Gamma_{\mathrm{m}}(-1)-\Gamma_{\mathrm{m}}(0)\}}{\Gamma_{\mathrm{m}}(1)-\Gamma_{\mathrm{m}}(-1)} \tag{4.9c}$$

Substituting these three terms in expression (4.8b) for Γ, we finally have

$$\Gamma = \frac{\{\Gamma_{\mathrm{m}}-\Gamma_{\mathrm{m}}(0)\}\{\Gamma_{\mathrm{m}}(1)-\Gamma_{\mathrm{m}}(-1)\}}{\{\Gamma_{\mathrm{m}}-\Gamma_{\mathrm{m}}(-1)\}\{\Gamma_{\mathrm{m}}(1)-\Gamma_{\mathrm{m}}(0)\}+\{\Gamma_{\mathrm{m}}-\Gamma_{\mathrm{m}}(1)\}\{\Gamma_{\mathrm{m}}(-1)-\Gamma_{\mathrm{m}}(0)\}} \tag{4.10}$$

Note 1

This procedure for arriving at the true value of the reflection coefficient of a one-port network is quite conventional. It is performed when the connection between the measurement plane and the one-port network is modelled as a passive two-port network resulting from the cascaded arrangement of connectors, adapters and cables. If we denote the matrix of a two-port network of this kind by $[E]$, we can see that the transmission terms appear simply as their product $E_{12}E_{21}$ which can be equated to E_{RF}.

Note 2

Manufacturers often express the error $\Delta\rho$ obtained in the modulus of the true reflection coefficient $\Gamma = \rho e^{j\theta}$ of a one-port network in the form of a second-degree polynomial in ρ which we can easily express by taking the difference between the measured coefficient and the true coefficient:

$$\Gamma_{\mathrm{m}}-\Gamma = E_{\mathrm{DF}}+\frac{E_{\mathrm{RF}}\Gamma}{1-E_{\mathrm{SF}}\Gamma}-\Gamma$$

Expanding $1/(1-E_{\mathrm{SF}}\Gamma)$ gives

$$\Gamma_{\mathrm{m}}-\Gamma = E_{\mathrm{DF}}+E_{\mathrm{RF}}\Gamma(1+E_{\mathrm{SF}}\Gamma+E_{\mathrm{SF}}{}^{2}\Gamma^{2}+\ldots)-\Gamma$$

We assume that the modulus of the source reflection coefficient is low enough to allow us to use a first-order expansion of the parentheses; hence

$$\Gamma_{\mathrm{m}}-\Gamma \approx E_{\mathrm{DF}}+(E_{\mathrm{RF}}-1)\Gamma+E_{\mathrm{RF}}E_{\mathrm{SF}}\Gamma^{2}$$

Taking the moduli, we obtain the inequality

$$\Delta\rho \leqslant |E_{\mathrm{DF}}|+|E_{\mathrm{RF}}-1|\rho+|E_{\mathrm{RF}}E_{\mathrm{SF}}|\rho^{2} \tag{4.11}$$

The equality corresponds to $\Delta\rho_{\max}$. The phase error $\Delta\theta$ satisfies the inequality obtained from inspection of Fig. 4.6:

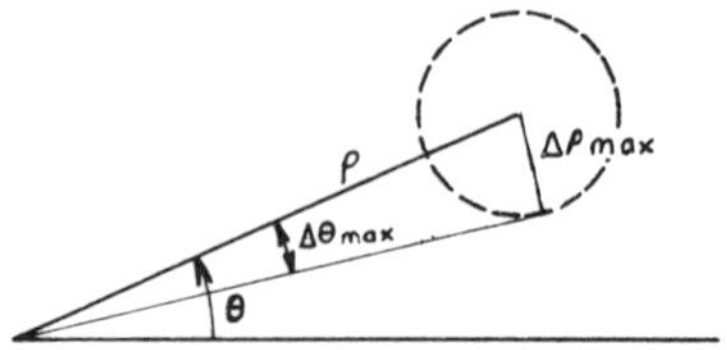

Fig. 4.6 Maximum phase error obtained when measuring a one-port network.

$$\Delta\theta \leqslant \arcsin\left(\frac{\Delta\rho_{\max}}{\rho}\right) \tag{4.12}$$

We still have to find the other three terms E_{LF}, E_{XF} and E_{TF}.

As E_{LF} represents a reflection coefficient, it is logical to attempt to obtain it by retaining the previous configuration giving S_{11M} and by directly connecting port 2 of the analyser to port 1. This implies that $S_{11} = S_{22} = 0$ and $S_{12} = S_{21} = 1$ and consequently leads to the flowchart shown in Fig. 4.7. Then making $\Gamma_m = E_{LFM}$ in Eqn (4.8b) gives

$$E_{LF} = \frac{E_{LFM} - E_{DF}}{E_{SF}(E_{LFM} - E_{DF}) + E_{RF}} \tag{4.13}$$

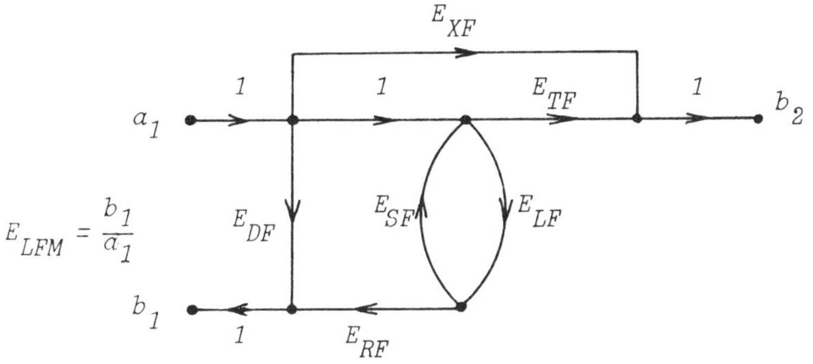

Fig. 4.7 *Measuring the reflection coefficient of the second analyser port.*

For E_{XF} ports 1 and 2 remain disconnected, $S_{12} = S_{21} = 0$ and we perform a transmission measurement. Inspection of the graph of Fig. 4.4(a) shows that no loop is then adjacent to the only path from a_1 to b_2, yielding the direct result

$$E_{XF} = E_{XFM} \tag{4.14}$$

For E_{TF} we again directly connect ports 1 and 2, but this time a transmission measurement is performed. Mason's rule applied to Fig. 4.7 gives

$$E_{TFM} = E_{XF} + \frac{E_{TF}}{1 - E_{SF}E_{LF}}$$

and hence

$$E_{TF} = (E_{TFM} - E_{XF})(1 - E_{SF}E_{LF}) \tag{4.15}$$

The six error terms have therefore been evaluated by using three precision loads on the one hand and by effecting three 'autocalibrations' on the other. As these preliminary operations are performed at each frequency for the 12 terms, we are then in a position to find the true scattering parameters of the two-port network under test using its four measured parameters. Expressions for the latter are in fact easily obtained by solving the graphs of Figs 4.4(a) and 4.4(b) using Mason's formula (Eqn (A1)).

The common denominator of S_{11M} and S_{21M} is

$$D_F = 1-(E_{SF}S_{11}+E_{LF}S_{22}+E_{SF}E_{LF}S_{12}S_{21})+E_{SF}E_{LF}S_{11}S_{22}$$

Introducing $\Delta = S_{11}S_{22}-S_{12}S_{21}$ we obtain

$$D_F = 1-E_{SF}S_{11}-E_{LF}S_{22}+E_{SF}E_{LF}\Delta$$

Thus

$$S_{11M} = E_{DF}+E_{RF}\frac{S_{11}-E_{LF}\Delta}{D_F} \tag{4.16a}$$

and

$$S_{21M} = E_{XF}+E_{TF}\frac{S_{21}}{D_F} \tag{4.16b}$$

However,

$$D_R = 1-E_{SR}S_{22}-E_{LR}S_{11}+E_{SR}E_{LR}\Delta$$

Hence

$$S_{22M} = E_{DR}+E_{RR}\frac{S_{22}-E_{LR}\Delta}{D_R} \tag{4.17a}$$

and

$$S_{12M} = E_{XR}+E_{TR}\frac{S_{12}}{D_R} \tag{4.17b}$$

Each of the true S parameters will therefore depend on the 12 error terms and the four measured coefficients. The inversion formulae of (4.16) and (4.17) will be found in the application notes for the 'new generation' network analysers (see the Bibliography). These recent systems are controlled by an internal computer which, incorporating the calibrations introduced at the start of measurement, reconstructs the true parameters selected virtually in real time.

4.2. Amplifier noise factor measurement

Determination of the noise factor of a cascade of amplifiers requires knowledge of the available power gain expression for each element in the chain. Moreover, the values measured are very often subject to errors such as that due to the variation in the reflection coefficient seen at the input during switching of the noise sources. Thus, before discussing the noise factor and how it is measured, we shall devote the following section to the various power gain expressions and mismatch terms which may be encountered in this field.

POWER GAIN AND MISMATCH EXPRESSIONS

The active two-port network inserted between impedances Z_1 and Z_2 can be redrawn as shown in Fig. 4.8.

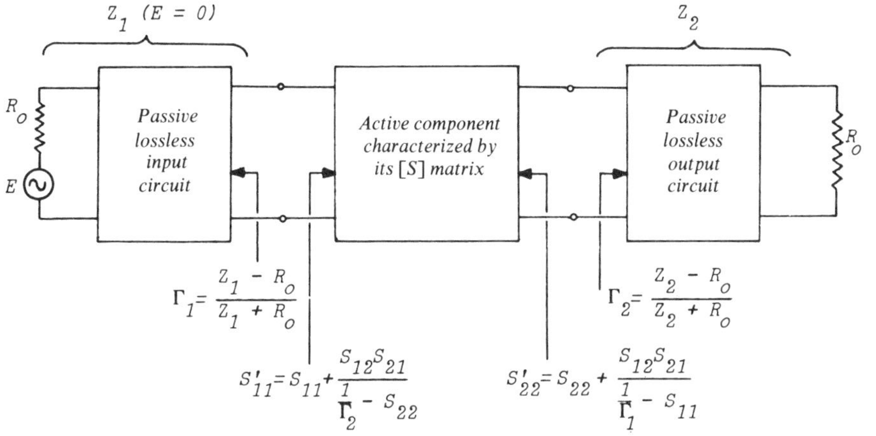

Fig. 4.8 Reflection coefficients seen at the ports of a two-port network.

The **forward Transducer power gain** G_T represents the ratio of the power dissipated in Z_2 to the power available from the source. Its expression, which was derived in Chapter 2, is

$$G_T = \frac{(1-|\Gamma_1|^2)|S_{21}|^2(1-|\Gamma_2|^2)}{|(1-\Gamma_1 S_{11})(1-\Gamma_2 S_{22})-\Gamma_1\Gamma_2 S_{12}S_{21}|^2}$$

We now define a quantity essential for calculating the noise factor, i.e. the **available power gain** G_A, which is equal to the ratio of the power available at the output to the power available at the input. G_A is therefore the transducer gain which we would obtain if the output were matched, i.e. for $\Gamma_2 = S'_{22}{}^*$. Now, the other expression for G_T in terms of S'_{22} was

$$G_T = \frac{(1-|\Gamma_1|^2)|S_{21}|^2(1-|\Gamma_2|^2)}{|1-\Gamma_1 S_{11}|^2|1-\Gamma_2 S'_{22}|^2}$$

We therefore obtain

$$G_A = \frac{1-|\Gamma_1|^2}{|1-\Gamma_1 S_{11}|^2(1-|S'_{22}|^2)}|S_{21}|^2 \tag{4.18}$$

The available power gain is therefore only a function of Γ_1 which may be dictated by certain constraints. If we wish to achieve the condition $F = F_{min}$ for a low noise amplifier it is necessary that $\Gamma_1 = \Gamma_{opt}$ (see Eqn (2.54)) and we introduce the **associated gain** G_a such that $G_a = G_A(\Gamma_1 = {}_{opt})$ or

$$G_a = \frac{1-|\Gamma_{opt}|^2}{|1-\Gamma_{opt} S_{11}|^2(1-|S'_{22}|^2)}|S_{21}|^2 \tag{4.19}$$

This is the quantity used in Eqn (2.58) which gives the minimum noise figure.

Finally, we define the **power gain** G which is the ratio of the power dissipated in the load Z_2 to the power delivered to the amplifier input. G is therefore the transducer gain corresponding to $\Gamma_1 = S'_{11}{}^*$. G_T can now be written in terms of S'_{11}:

$$G_T = \frac{(1-|\Gamma_1|^2)|S_{21}|^2(1-|\Gamma_2|^2)}{|1-\Gamma_1 S'_{11}|^2|1-\Gamma_2 S_{22}|^2}$$

and hence

$$G = |S_{21}|^2 \frac{1-|\Gamma_2|^2}{(1-|S'_{11}|^2)|1-\Gamma_2 S_{22}|^2} \tag{4.20}$$

We are now in a position to quantify the input and output mismatch.

The **input mismatch** M_1 is the ratio of the power delivered to the input to the power available at the input, such that

$$G_T = M_1 G$$

and hence

$$M_1 = \frac{(1-|\Gamma_1|^2)(1-|S'_{11}|^2)}{|1-\Gamma_1 S'_{11}|^2} \tag{4.21}$$

The **output mismatch** M_2 is the ratio of the power dissipated in the load to the power available at the output, which gives the relation

$$G_T = G_A M_2$$

where

$$M_2 = \frac{(1-|\Gamma_2|^2)(1-|S'_{22}|^2)}{|1-\Gamma_2 S'_{22}|^2} \tag{4.22}$$

NOISE FACTOR EXPRESSIONS

It is useful to regard these various expressions as deriving from the same basic definition quantifying the degradation of the signal-to-noise ratio through a two-port network. The actual measurements are based on more restrictive formulations involving the concept of equivalent noise temperature.

Signal-to-noise ratio

Let $(S/N)_{in} = S_{in}/N_{in}$ be the ratio of signal power to noise power at the amplifier input. The signal power S_{in} delivered to the input is given by

$$S_{in} = \frac{E_g^2}{2} \frac{R_{in}}{|Z_g + Z_{in}|^2}$$

and the noise N_{in} is given by

$$N_{in} = \overline{v_g^2} \frac{R_{in}}{|Z_g + Z_{in}|^2}$$

where Z_{in} is the input impedance of the amplifier and Z_g is the internal impedance of a Thévenin generator with electromotive force E_g connected as shown in Fig. 4.9. Thus

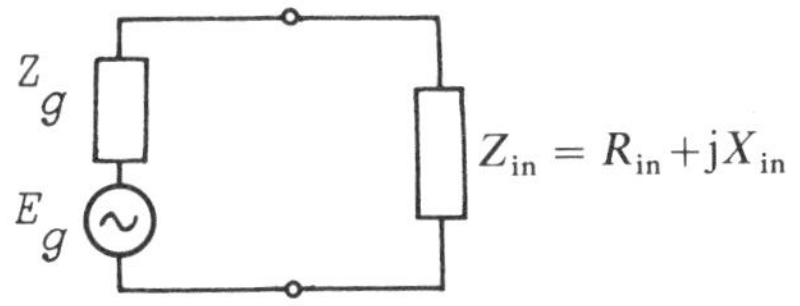

Fig. 4.9 *Thévenin generator at the amplifier input.*

$$\left(\frac{S}{N}\right)_{in} = \frac{1}{2}\frac{E_g^{\,2}}{v_g^{\,2}} = \frac{1}{2}\frac{E_g^{\,2}}{4kT\mathrm{Re}(Z_g)\Delta f}$$

The signal-to-noise ratio is therefore independent of the input impedance and depends only on the source. It is often rewritten in the form

$$\frac{S_{in}}{N_{in}} = \frac{E_g^{\,2}/8\mathrm{Re}(Z_g)}{kT\Delta f}$$

which means that the signal-to-noise ratio at the amplifier input is equal to the ratio of the available signal power at the source terminals to the available noise power at these same terminals. We should not conclude, however, that this only applies to the matching condition.

As the amplifier has a certain intrinsic noise, the signal-to-noise ratio at the output is subject to degradation: $(S/N)_{out}$ is less than $(S/N)_{in}$ and is independent of the load impedance Z_L. The noise factor F is thus defined quite naturally by the ratio

$$F = \frac{(S/N)_{in}}{(S/N)_{out}} \tag{4.23}$$

This fundamental definition is independent of both input and output matching of the two-port network; no assumptions are made concerning the gain.

Expressions derived from the definition of the noise factor

We first derive an expression based on the concept of the additional noise power N_a of the amplifier presented at the input. If G is the power gain at the signal frequency, we can write $N_{out} = G(N_{in} + N_a)$ and $S_{out} = GS_{in}$. Hence

$$\left(\frac{S}{N}\right)_{out} = \frac{S_{out}}{N_{out}} = \frac{S_{in}}{N_{in} + N_a}$$

The gain term only appears at an intermediate stage. The noise factor is therefore expressed solely in terms of N_{in} and N_a:

$$F = 1 + \frac{N_a}{N_{in}} \tag{4.24a}$$

We note that the noise factor takes the form of a ratio which is independent of the input impedance Z_{in} and therefore of the load impedance Z_L as we were able to show for the noisy two-port network model. The noise factor is calculated by presenting the noise sources $\overline{v^2}$ and $\overline{i^2}$ to the input, generally with a certain correlation coefficient. This results in an expression which is purely a function of the source impedance Z_g or, in accordance with Eqn (2.43),

$$F = 1 + \frac{\overline{(v + Z_g i)(v + Z_g i)^*}}{\overline{v_g^2}}$$

We can express this more concisely as

$$F = 1 + \frac{\overline{v_{in}^2}}{\overline{v_g^2}} \tag{4.24b}$$

We now derive an expression based on the concept of excess noise power N_r added by the amplifier at the output. Note that $N_r = GN_a$, and if we put $(N_{out})_O = GN_{in}$ we obtain

$$F = \frac{N_{out}}{(N_{out})_O} = 1 + \frac{N_r}{(N_{out})_O} \tag{4.25a}$$

The noise factor appears as the ratio of the output noise power of the real amplifier to the output noise power of the noiseless amplifier. The second equality is used in the calculations when all the noise sources are presented at the output of the two-port network. This leads to the Thévenin representation shown in Fig. 4.10 where Z_{out} is the output impedance. The ratio $N_r/(N_{out})_O$ can be simplified by introducing the proportionality coefficient $R_L/|Z_{out} + Z_L|^2$ and we thus obtain the counterpart of formula (4.24b):

$$F = 1 + \frac{\overline{v_{out}^2}}{\overline{v_{g\,out}^2}} \tag{4.25b}$$

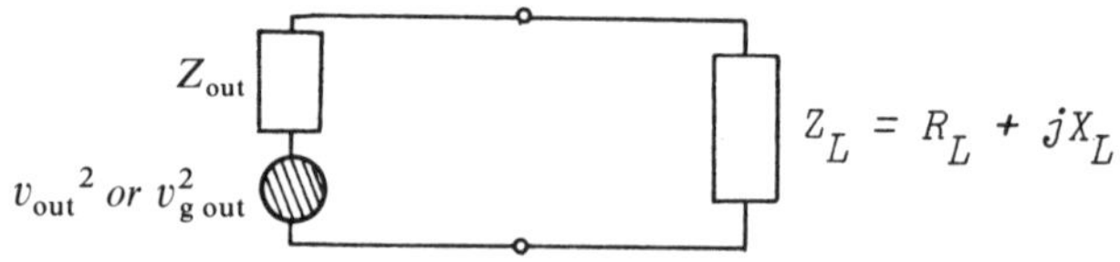

Fig. 4.10 Thévenin representation at the output.

Formulae obtained under matching conditions

These formulae, on which actual measurements are often based, imply that matching is achieved at both the input and the output. Thus, according to definitions (4.24a) and (4.25a), the noise powers involved are available powers which can be rewritten as follows:

$$N_{in} = kT\Delta f$$

$$(N_{out})_0 = G_A{}^\circ N_{in} = G_A{}^\circ kT\Delta f$$

$$N_{out} = (N_{out})_0 + N_r = FG_A{}^\circ kT\Delta f$$

These equalities imply

$$N_r = (F-1)G_A{}^\circ kT\Delta f$$

and

$$N_a = (F-1)kT\Delta f \tag{4.26}$$

Note

The quantity $G_A{}^\circ\Delta f$ is equal to the product of the maximum available power gain of the amplifier and the power bandwidth Δf such that

$$G_A{}^\circ\Delta f \equiv \int_0^\infty G_A(f)\,df$$

Such expressions are particularly suitable for analysing the noise factors of cascaded two-port networks. For example, if it is assumed that the matching problems have been resolved, the output noise powers for two two-port networks 1 and 2 are respectively

$$(N_{out})_1 = F_1(G_A{}^\circ)_1 kT\Delta f$$

$$(N_{out})_2 = (G_A{}^\circ)_2(N_{out})_1 + (N_r)_2$$

or

$$(N_{out})_2 = (G_A{}^\circ)_1(G_A{}^\circ)_2 F_1 kT\Delta f + (F_2-1)(G_A{}^\circ)_2 kT\Delta f$$

By equating with

$$(N_{out})_2 = F(G_A{}^\circ)_1(G_A{}^\circ)_2 kT\Delta f$$

we obtain the conventional expression

$$F = F_1 + \frac{F_2 - 1}{(G_A{}^\circ)_1} \tag{4.27}$$

of which Friis' formula is the generalization to n two-port networks at identical temperatures. We note that F_2 is the noise factor of two-port network 2 driven by a Thévenin source whose internal impedance is the output impedance of two-port network 1.

Noise factor and equivalent noise temperature

We can also use the 'additional noise temperature' T_A which is defined in terms of the noise factor. For this purpose, we assume that the system under examination is characterized with respect to a standard reference temperature $T_0 = 290\,\text{K}$. The available input noise power is then $N_{in} = kT_0\Delta f$ and by analogy we write

$$N_a = kT_A\Delta f$$

which, because of the relation

$$N_a = (F-1)kT_0\Delta f$$

gives

$$T_A = (F-1)T_0 \tag{4.28a}$$

and therefore

$$F = \frac{T_0 + T_A}{T_0} \tag{4.28b}$$

The advantage of this equivalent temperature concept is that it allows us to measure the noise of a device without recourse to a reference temperature. In fact, the available output noise power at any temperature T is

$$N_{out} = k(T + T_A)G_A{}^{\circ}\Delta f$$

N_{out} is therefore a linear function of T and the representative straight line will intersect the axis of the abscissa at $-T_A$ as shown in Fig. 4.11.

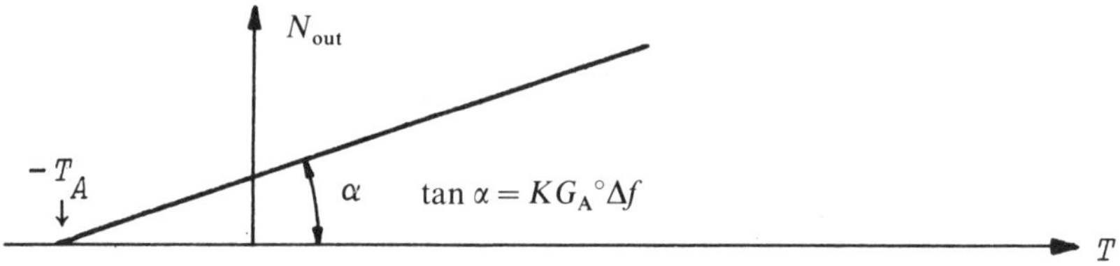

Fig. 4.11 Graphical determination of the equivalent noise temperature.

Modern low noise amplifiers have temperatures which may be much lower than T_0 and they are often associated with receiving antennas producing temperatures of about 40 K or less because they point at geostationary satellite orbits; for these two reasons, the noise factor concept is abandoned in such cases.

METHODS OF MEASURING THE NOISE FACTOR

After discussing the historical 3 dB method, based on the use of a noise diode, which has had to be abandoned because of its limitations at high frequencies owing to the electron transit time, we shall describe the principles of an automatic measurement using only a single noise source. However, manual measurements using two hot/cold standard sources are still the most accurate. It should be noted that the methods considered below give *a priori* an average noise factor, but a point-by-point measurement, say in a band of a few megahertz, may also be needed, and this is possible with recent equipment.

3 dB method

Let us give an example using the noise factor expression (4.24b) or

$$F = \frac{\overline{v_g^2} + \overline{v_{in}^2}}{\overline{v_g^2}}$$

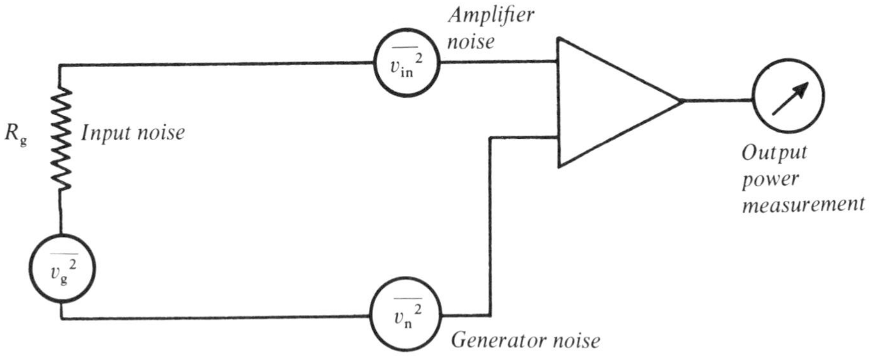

Fig. 4.12 *Noise measurement using the 3 dB method.*

We connect the noise generator to the amplifier input as shown in the schematic diagram in Fig. 4.12. When the generator is off, the noise power reading at the output will be proportional to $\overline{v_{in}^2} + \overline{v_g^2}$, and when it is on it will be proportional to $\overline{v_{in}^2} + \overline{v_g^2} + \overline{v_n^2}$. The measuring principle consists of regulating the noise level generated by the equipment in such a way as to double the reading at the output; we shall then have

$$\overline{v_n^2} = \overline{v_{in}^2} + \overline{v_g^2}$$

and the noise factor can be written in the form

$$F = \frac{\overline{v_n^2}}{\overline{v_g^2}}$$

Now, v_g is a thermal noise and consequently

$$\overline{v_g^2} = 4kR_gT\Delta f$$

The generator itself produces a shot noise generated by a noise diode which must be operating under saturation conditions if we wish to use the Schottky formula without correction. If the mean current output of the diode is I, the mean quadratic value of the associated noise current i_n is given by

$$\overline{i_n^2} = 2qI\Delta f$$

where q is the elementary charge and is equal to 1.602×10^{-19} C. If the generator presents a resistance R, the available noise power will be

$$\frac{\overline{v_n^2}}{R} = R\overline{i_n^2}$$

Hence

$$\overline{v_n^2} = 2qIR^2\Delta f$$

In fact, we connect the equipment directly to the amplifier input and therefore R and R_g are identical; thus

$$F = \frac{qR_g}{2kT} I \tag{4.29a}$$

For $R_g = 50\,\Omega$ and an ambient temperature of 290 K ($T = T_0 = 290$ K), the coefficient $qR_g/2kT$ is equal to 1000; hence

$$F_{dB} = 30 + 10 \log I \tag{4.29b}$$

In practice the power meter, which must have a passband higher than the power bandwidth Δf of the amplifier, is simply used as a zero indicator. In the second measurement phase, we prefer to use a 3 dB calibrated attenuator.

Automatic measurement

We can use the formula

$$F = \frac{(S/N)_{in}}{(S/N)_{out}} = \frac{S_{in}/N_{in}}{S_{out}/N_{out}}$$

and if, at the output, we cannot distinguish the measure of the signal S_{out} from the measure of the noise N_{out} we put $P_{out} = S_{out} + N_{out}$ and write

$$F = \frac{S_{in}/N_{in}}{P_{out}/N_{out} - 1}$$

Effectively, the signal here will be the noise produced by the source associated with the equipment. The ratio S_{in}/N_{in} is therefore known; it is a constant K_1 of the device. By using a multivibrator to control the periodic turn-on and turn-off of the noise source, we obtain the values of P_{out} and N_{out} sequentially. In fact, if an automatic gain control device makes $P_{out} = K_2$, the noise factor will be simply a function of N_{out}. The measuring equipment must be capable of calculating the quantity

$$F_{dB} = 10 \log\left(\frac{K_1}{K_2/N_{out} - 1}\right) = 10 \log K_1 - 10 \log\left(\frac{K_2}{N_{out}} - 1\right)$$

Expressions for F derived from measurements of the noise temperature T_A

This time we use the relation $F = 1 + T_A/T_0$. Suppose that we have two sources delivering available powers $N_{in}' = kT'\Delta f$ and $N_{in}'' = kT''\Delta f$ where T' and T'' are the temperatures associated with the sources. Connecting these noise generators in turn to the amplifier input and assuming that the matching condition has been achieved in both cases, we obtain the available output powers

$$N_{out}' = kT'G_A^\circ\Delta f + kT_AG_A^\circ\Delta f$$

and

$$N_{out}'' = kT''G_A^\circ\Delta f + kT_AG_A^\circ\Delta f$$

respectively.

For the first measurement the value N_{out}' will correspond to a particular reading on the level detector (the power meter) connected at the amplifier output. If we assume that T'' is greater than T' it will be necessary to attenuate by y dB if we wish to obtain the same reading for the second measurement. We therefore infer that

$$y = \frac{N_{out}''}{N_{out}'} = \frac{T'' + T_A}{T' + T_A}$$

Hence

$$T_A = \frac{T'' - yT'}{y - 1} \tag{4.30}$$

and consequently

$$F = 1 + \frac{T''/T_0 - yT'/T_0}{y - 1} \tag{4.31}$$

The system described above is shown in the block diagram in Fig. 4.13.

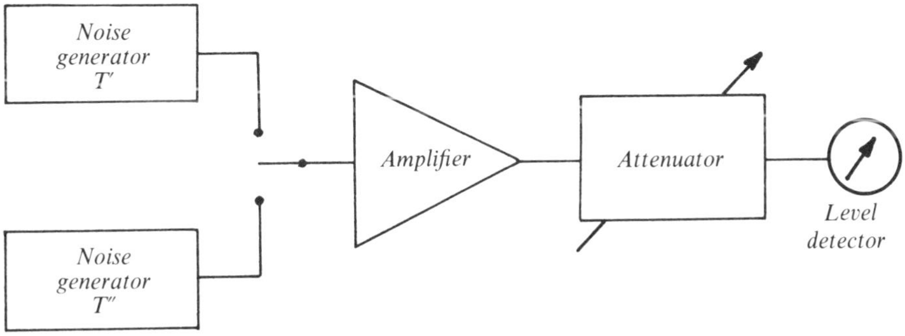

Fig. 4.13 Noise factor measurement using the two-source method.

If we use resistive terminations immersed in boiling liquids at a given pressure as sources (these are defined as standard sources) we have a means of evaluating F with minimum error. We use liquid nitrogen to obtain the standard cold source with $T' = T_{\text{cold}} = 77.3\,\text{K}$. The standard hot source is obtained by keeping the termination in an oven stabilized at the temperature of boiling water so that $T'' = T_{\text{hot}} = 373.2\,\text{K}$.

In practice, we only use a single noise source, i.e. $T' = T_0$ (termination at ambient temperature) and we write $T'' = T_0 + T_R$ where T_R is the excess noise temperature produced by the generator. The ratio T_R/T_0 is known as the **excess noise ratio** ENR and is generally expressed in decibels:

$$(\text{ENR})_{\text{dB}} = 10\log\left(\frac{T_R}{T_0}\right) \tag{4.32a}$$

From the expression for F as a function of T'' and T' we therefore obtain

$$F_{\text{dB}} = (\text{ENR})_{\text{dB}} - 10\log(y-1) \tag{4.32b}$$

We note that Eqn (4.31) allows us to perform a comparatively simple error calculation leading to ΔF or $\Delta F/F$. In fact, the mathematical relation

$$\Delta F = \left|\frac{\partial F}{\partial T''}\right|\Delta T'' + \left|\frac{\partial F}{\partial T'}\right|\Delta T' + \left|\frac{\partial F}{\partial y}\right|\Delta y$$

gives

$$\Delta F = \frac{1}{T_0(y-1)}\Delta T'' + \frac{y}{T_0(y-1)}\Delta T' + \frac{T''-T'}{T_0(y-1)^2}\Delta y$$

and, assuming once again that $T' = T_0$ and $T'' = T_0 + T_R$, we note that

$$\frac{\Delta T''}{T_0} = \frac{\Delta T_R}{T_0} = \Delta(\text{ENR})$$

Hence

$$\Delta F = \frac{\Delta(\text{ENR})}{y-1} + \frac{\text{ENR}}{(y-1)^2}\Delta y$$

and as a relative value

$$\frac{\Delta F}{F} = \frac{\Delta(\text{ENR})}{\text{ENR}} + \frac{\Delta y}{y-1}$$

The error due to the source and that due to the attenuation are separate in this equation. Taking into account the equality

$$\text{d}F_{\text{dB}} = \text{d}(10\log F) = (10\log \text{e})\frac{\text{d}F}{F}$$

we can write in decibels ($10\log \text{e} \approx 4.34$)

$$\Delta F_{\text{dB}} = \Delta(\text{ENR})_{\text{dB}} + 4.34\frac{\Delta y}{y-1} \tag{4.33}$$

The dominant error term is that corresponding to the ENR. The ENR of gas discharge tubes comprising an ionized inert gas and of solid state sources using an avalanche noise diode is approximately 15–16 dB (with an industrial accuracy to ± 0.5 dB). It is only by means of direct calibration using hot/cold standard sources that it is possible to obtain a figure of 0.10 dB in the frequency bands below 18 GHz.

For comparison, we quantify the accuracy of a noise factor measured using standard sources. We take $\Delta T_{\text{hot}} = \Delta T_{\text{cold}} = 1$ K and $\Delta y_{\text{dB}} = 0.03$. For $F = 3$ dB, we obtain $y = 1.805$ and hence $\Delta y = 0.013$; consequently, $\Delta F = 0.012$ (source) $+ 0.020$ (attenuator) $= 0.032$, or $F = 2 \pm 0.032$, and therefore $T_A = 290 \pm 9$ K or $F = 3 \pm 0.07$ dB.

The error with respect to the sources is somewhat lower than that obtained using the precision attenuator; however, it is advisable to take the two following points into consideration: firstly, using two sources poses the difficult problem of the variation of the reflection coefficient at the amplifier input; secondly, it is extremely difficult to determine the temperature of the source–amplifier link.

4.3. Characterization of the spectral purity of oscillators

Owing to the various types of noise (thermal, Schottky, flicker etc.) which affect the active component in particular, the band emitted by the oscillator undergoes some broadening. Both amplitude and phase modulation are present. For a good oscillator, we can legitimately disregard the amplitude noise; thus our analysis will concentrate on the phase noise which can be examined either in the time domain or in the frequency domain. We shall begin with the latter, after first recalling the concept of frequency modulation noise which is the theoretical basis of the classical measurement of $\mathscr{L}(f_m)$ using a spectrum analyser.

FREQUENCY MODULATION NOISE

In this type of modelling we assume that a carrier f_0 undergoes frequency modulation due to noise. If Δf is the frequency excursion and f_m is the modulating frequency we have

$$f = f_0 + \Delta f \cos(2\pi f_m t)$$

The voltage produced by an ideal oscillator modulated in this way is

$$v_m(t) = V_0 \sin\{2\pi f_0 t + m \sin(2\pi f_m t)\}$$

where $m = \Delta f / f_m$ is the modulation factor. Expanding $v_m(t)$ gives

$$v_m(t) = V_0[\sin(2\pi f_0 t) \cos\{m \sin(2\pi f_m t)\} + \cos(2\pi f_0 t) \sin\{m \sin(2\pi f_m t)\}]$$

Introducing the Bessel functions $J_n(m)$ we obtain

$$\frac{v_m(t)}{V_0} = J_0(m)\sin(2\pi f_0 t) + \sum_{n=1}^{\infty} J_n(m)[\sin\{2\pi(f_0 + nf_m)t\} + (-1)^n \sin\{2\pi(f_0 - nf_m)t\}]$$

Since $J_{-n}(m) = (-1)^n J_n(m)$, this finally gives

$$v_m(t) = V_0 \sum_{n=-\infty}^{\infty} J_n(m)\sin\{2\pi(f_0 + nf_m)t\} \qquad (4.34)$$

Now

$$\sum_{n=-\infty}^{\infty} J_n^2(m) = 1$$

and the power transferred by the frequency-modulated carrier is the same as that transferred without modulation.

However, for small m we have the approximation

$$J_n(m) \approx \frac{1}{n!}\left(\frac{m}{2}\right)^n \qquad (n>0)$$

and for $n = 0$

$$J_0(m) \approx 1 - \left(\frac{m}{2}\right)^2$$

Thus a first-order expansion of Eqn (4.34) gives

$$v_m(t) \approx V_0 \sin(2\pi f_0 t) + \frac{mV_0}{2}\sin\{2\pi(f_0 + f_m)t\} - \frac{mV_0}{2}\sin\{2\pi(f_0 - f_m)t\}$$

The positive part of the amplitude spectrum $S_v(f)$ obtained by means of a Fourier transform is shown in Fig. 4.14. As V_0 is a peak value, taking into account the negative part of the spectrum, for a load of 1 Ω we have power P_0 on the carrier,

$$P_0 = \frac{1}{2} \times 2\frac{V_0^2}{4} = \frac{V_0^2}{4}$$

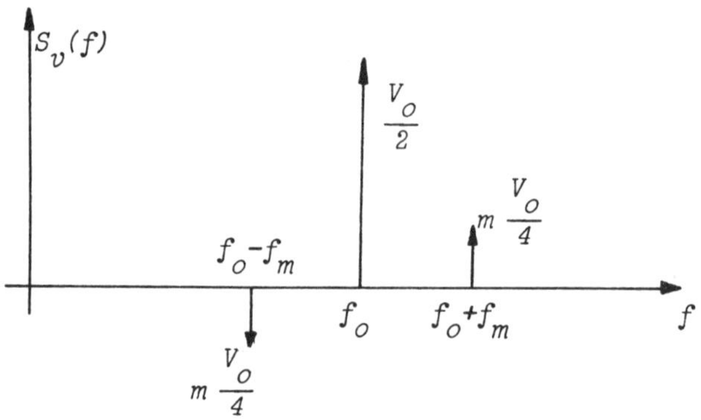

Fig. 4.14 *Amplitude spectrum.*

and power P_{fm} on the band f_0+f_m,

$$P_{fm} = \frac{1}{2} \times 2\frac{m^2 V_0^2}{16} = \frac{m^2 V_0^2}{16}$$

Hence

$$\frac{P_{fm}}{P_0} = \frac{1}{4}\left(\frac{\Delta f}{f_m}\right)^2 \tag{4.35}$$

P_{fm} corresponds to the noise power in a 1 Hz band centred at a distance f_m from the carrier. The ratio P_{fm}/P_0 can therefore be measured using a spectrum analyser for example. Such an instrument can be used only if its local oscillator is of a much better quality than the oscillator under test. However, the filter used is often gaussian and its power bandwidth B, which is certainly higher than 1 Hz, makes it necessary to subtract $10 \log B$ from the measurement of P_{fm} (in decibels). The value obtained after correction is denoted by $\mathscr{L}(f_m)$ which is expressed in decibels below the carrier per hertz by

$$\mathscr{L}(f_m) = \frac{P_{fm}\,(\text{in 1 Hz})}{P_0} \tag{4.36}$$

CHARACTERIZATION IN THE FREQUENCY DOMAIN

If we consider only the phase noise $\phi(t)$, the voltage $v(t)$ of amplitude V_0 produced by an oscillator at frequency f_0 is given by

$$v(t) = V_0 \sin\{2\pi f_0 t + \phi(t)\}$$

Since $\phi(t)$ is a random process which is assumed to be stationary, its autocorrelation function Γ_ϕ can be written in the form

$$\Gamma_\phi(\tau) = E\{\phi(t)\phi(t+\tau)\}$$

This second-order mathematical expectation represents a statistical mean equal to the temporal mean calculated in the case of ergodicity. The following equalities are derived from this assumption:

$$\Gamma_\phi(\tau) = E\{\phi(t)\phi(t+\tau)\} = \overline{\phi(t)\phi(t+\tau)}$$

with

$$\overline{\phi(t)\phi(t+\tau)} = \lim_{T\to\infty} \frac{1}{2T}\int_{-T}^{T} \phi(t)\phi(t+\tau)\,dt$$

We note that for zero argument τ the autocorrelation function is the average power of the process:

$$\Gamma_\phi(0) = E\{\phi^2(t)\} = \overline{\phi^2(t)}$$

The transition to the frequency domain is effected by taking the Fourier transform of $\Gamma_\phi(\tau)$. We denote by $G_\phi(f)$ the function thus obtained:

$$G_\phi(f) = \mathrm{FT}\{\Gamma_\phi(\tau)\}$$

In accordance with the Wiener–Kintchine theorem, this function is the spectral power density of the process $\phi(t)$, which is expressed here in radians squared per hertz. Experimentally, we obtain this phase fluctuation spectral density by measuring the mean power of $\phi(t)$ in a 1 Hz window centred on $f_0 + f_m$, where f_m is the frequency deviation with respect to the carrier; hence

$$G_\phi(f) = \frac{\phi_{\mathrm{rms}}{}^2(f_m)}{1\ \mathrm{Hz}}\ \mathrm{rad}^2\ \mathrm{Hz}^{-1}$$

In fact, it is necessary to divide by the width B of the window.

This measurement can be performed using a base band spectrum analyser; again, it is necessary to find $\phi(t)$. Theoretically, this is simple if we have a reference oscillator producing a voltage $v_r(t) = V_{0r}\cos(2\pi f_0 t)$.

When $v(t)$ and $v_r(t)$ are applied to a multiplier, the output signal $u(t)$ will be proportional to $\sin\{2\pi f_0 t + \phi(t)\}\cos\{2\pi f_0(t)\}$, or

$$u(t) = K[\sin\{2\pi 2 f_0 t + \phi(t)\} + \sin\{\phi(t)\}]$$

After filtering the frequency $2f_0$, we obtain

$$u_1(t) = K_1 \sin\{\phi(t)\}$$

at the low pass filter output. For a good oscillator satisfying

$$\left|\frac{\mathrm{d}\phi(t)}{\mathrm{d}t}\right| \ll 2\pi f_0$$

we make the approximation $u_1(t) \approx K_1\phi(t)$.

In practice, the quadrature phase control can be established by using a phase-locked loop as shown in Fig. 4.15.

Returning now to the phase term $2\pi f_0 t + \phi(t)$ of the oscillator, we find that its instantaneous frequency $f(t)$ can be expressed by

$$f(t) = f_0 + \frac{1}{2\pi}\frac{\mathrm{d}\phi(t)}{\mathrm{d}t}$$

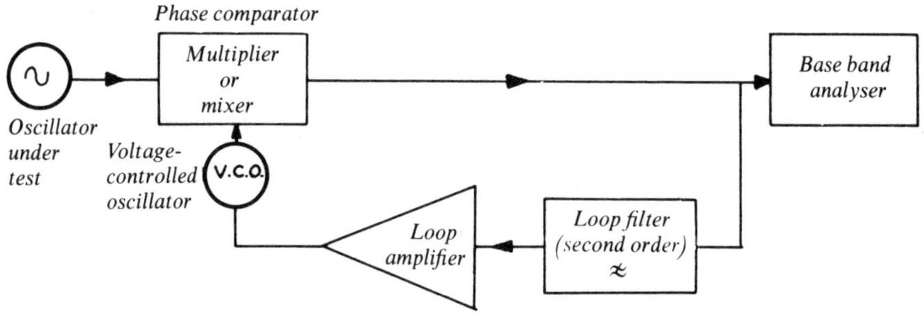

Fig. 4.15 *Measurement of the spectral density of the phase noise.*

but we are more interested in the relative frequency variation or the normalized instantaneous frequency deviation $y(t)$:

$$y(t) = \frac{f(t) - f_0}{f_0}$$

or

$$y(t) = \frac{1}{2\pi f_0} \frac{\mathrm{d}\phi(t)}{\mathrm{d}t} \tag{4.37}$$

We can regard $y(t)$ as the process resulting from the passage of $\phi(t)$ through a filter of response $(1/2\pi f_0)\,\mathrm{d}/\mathrm{d}t$ and therefore with a transfer function $\mathrm{j}2\pi f/2\pi f_0$. Thus the output power spectral density is

$$G_Y(f) = \frac{1}{4\pi^2 f_0^{\,2}} |\mathrm{j}2\pi f|^2 G_\phi(f)$$

The transition relation between the phase and frequency variation spectra is

$$f_0^{\,2} G_Y(f_m) = f_m^{\,2} G_\phi(f_m) \tag{4.38}$$

It is also a simple matter to establish a relationship with $\mathscr{L}(f_m)$. For a good oscillator, we have seen (Eqns (4.35) and (4.36)) that we can write

$$\mathscr{L}(f_m) = \frac{1}{4}\left(\frac{\Delta f}{f_m}\right)^2$$

Now $\Delta f/f_m$ represents a peak phase excursion $\Delta\varphi$ which is a function of the frequency deviation f_m; hence

$$\mathscr{L}(f_m) = \frac{1}{4}\{\Delta\varphi(f_m)\}^2 = \frac{1}{2}\{\Delta\varphi_{rms}(f_m)\}^2$$

Equating $\Delta\varphi_{rms}$ and ϕ_{rms}, we obtain the important result

$$\mathscr{L}(f_m) = \frac{1}{2} G_\phi(f_m) = \frac{1}{2} \frac{f_0^{\,2}}{f_m^{\,2}} G_Y(f_m) \tag{4.39}$$

CHARACTERIZATION IN THE TIME DOMAIN

We are basically attempting to arrive at the quantity $\sigma_y(\tau)$ equal to the square root of the variance of the relative frequency deviation y measured after a time τ. Still assuming ergodicity, we have

$$\sigma_y^{\,2}(\tau) = \overline{\{y(\tau) - \overline{y(\tau)}\}^2} = \overline{y^2(\tau)} - \{\overline{y(\tau)}\}^2$$

However, a measurement performed during the period t_k, $t_k + \tau$ will only yield a mean value $\bar{y}_k(t_k, \tau)$, denoted by $\bar{y}_k$, such that

$$\bar{y}_k = \frac{1}{\tau}\int_{t_k}^{t_k+\tau} y(t)\,\mathrm{d}t = \frac{1}{2\pi f_0 \tau}\{\phi(t_k+\tau) - \phi(t_k)\}$$

In practice, a counter is used to perform such measurements every T seconds over a period τ. The dead time t_d is therefore equal to $T-\tau$ as shown along the following time axis:

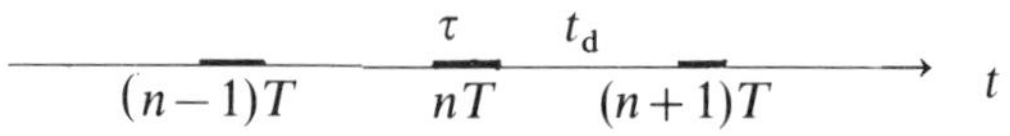

After N measurements, we can estimate the variance $\sigma_y^2(\tau)$ from the quantity

$$\sigma_{\bar{y}}^2(N, T, \tau) = \frac{1}{N}\sum_{n=1}^{N}\left(\bar{y}_n - \frac{1}{N}\sum_{n=1}^{N}\bar{y}_n\right)^2 = \frac{1}{N}\sum_{n=1}^{N}\bar{y}_n^{\,2} - \left(\frac{1}{N}\sum_{n=1}^{N}\bar{y}_n\right)^2$$

where

$$\bar{y}_n = \frac{1}{\tau}\int_{nT}^{nT+\tau} y(t)\,\mathrm{d}t = \frac{1}{2\pi f_0 \tau}\{\phi(nT+\tau)-\phi(nT)\} = \frac{1}{2\pi f_0}\,\frac{\Delta\phi_n}{\tau}$$

It might be expected that this estimator will tend to the true variance as N tends to infinity. In fact there is no convergence. $\sigma_{\bar{y}}^2(N, T, \tau)$ must be regarded as a random variable, obtained over the measuring period NT, and we can obtain its statistical mean by carrying out M measuring cycles.

D. W. Allan and J. A. Barnes have shown that the expression

$$\frac{1}{M}\sum_{i=1}^{M}(\sigma_{\bar{y}}^{\,2})_{(i)}(N, T, \tau)$$

generally approaches a limit as M tends to infinity. Thus the frequency instability is often defined by $\overline{\sigma_{\bar{y}}^{\,2}(N, T, \tau)}$ where

$$\overline{\sigma_{\bar{y}}^2(N, T, \tau)} = \lim_{M\to\infty}\frac{1}{M}\sum_{i=1}^{M}(\sigma_{\bar{y}}^2)_{(i)}\,(N, T, \tau) \qquad (4.40)$$

However, the second term, which is known as the **Allan variance**, is subject to a bias; this is not surprising since the mean value of $\bar{y}_n$ can only be estimated. Referring to the theory concerning the quality of estimators, we expect to eliminate the bias by putting

$$\sigma_{\bar{y}}^2(N, T, \tau) = \frac{1}{N-1}\sum_{n=1}^{N}\left(\bar{y}_n - \frac{1}{N}\sum_{n=1}^{N}\bar{y}_n\right)^2$$

Allan has shown that when T is very much greater than τ we effectively obtain the true variance.

ADVANTAGES OF TIME MEASUREMENTS

These are based on the fact that it is mathematically possible to move from the estimation of $\sigma_y(\tau)$ to $\mathscr{L}(f_m)$, as demonstrated in Appendix D. We can see that it will be necessary to use a fast Fourier transform (FFT) program. We take

N values of $\sigma_y(\tau)$ regularly spaced from $\tau_{\min}$ to $\tau_{\max}$; the incremental step is therefore

$$\Delta\tau = \frac{\tau_{\max} - \tau_{\min}}{N-1}$$

Application of the FFT to these N samples in accordance with the formula given in Appendix D, i.e.

$$\mathrm{FT}\{(2\pi f_0\tau)^2 \frac{1}{M}\sum_{k=1}^{M} (\sigma_y{}^2)_{(k)}'(N, T, \tau)\} = -2\left\{1 - \frac{1}{N^2}\,\frac{1-\cos(2\pi f N T)}{1-\cos(2\pi f T)}\right\} G_\phi(f)$$

will yield N values of $\mathscr{L}(f_m)$ between $f_{\min}$ and $f_{\max}$ with a constant increment Δf. If we define the relations

$$\Delta f = \frac{1}{\tau_{\max}}$$

$$f_{\max} = \frac{1}{\Delta\tau} = \frac{N-1}{\tau_{\max} - \tau_{\min}}$$

$$f_{\min} = f_{\max} - \frac{N-1}{\tau_{\max}} = \frac{\tau_{\min}}{\tau_{\max}} f_{\max}$$

it is clear that the time domain measurements enable us to obtain $\mathscr{L}(f_m)$ near the carrier. We shall illustrate this by the following example. We want to know $\mathscr{L}(f_m)$ between $f_{\min} = 1$ Hz and $f_{\max} = 100$ Hz. We therefore have $\tau_{\max}/\tau_{\min} = 100$; taking $N = 1000$, we examine $\sigma_y(\tau)$ between $\tau_{\min} = 0.1$ s and $\tau_{\max} = 10$ s. For $f_{\min} = 10$ Hz and $f_{\max} = 1000$ Hz, $\sigma_y(\tau)$ must be investigated between $\tau_{\min} = 0.01$ s and $\tau_{\max} = 1$ s. Figure 4.16 shows the former case.

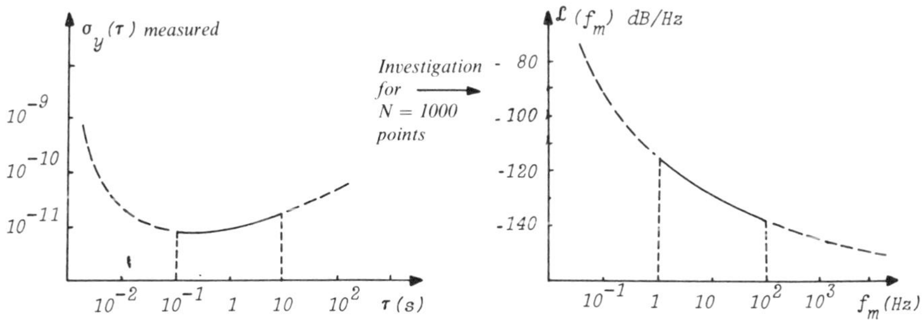

Fig. 4.16 *Correspondence between the time and frequency domains.*

Appendix A

Flowcharts

Solution by Mason's rule: application to voltage gain calculations

The amplifier, characterized by its $[S]$ matrix, is inserted between the source and load impedances Z_1 and Z_2 for which we derive the reflection coefficients

$$\Gamma_1 = \frac{Z_1 - R_0}{Z_1 + R_0} \quad \text{and} \quad \Gamma_2 = \frac{Z_2 - R_0}{Z_2 + R_0}$$

respectively. The corresponding schematic diagram is given in Fig. A1 and the flowchart in Fig. A2.

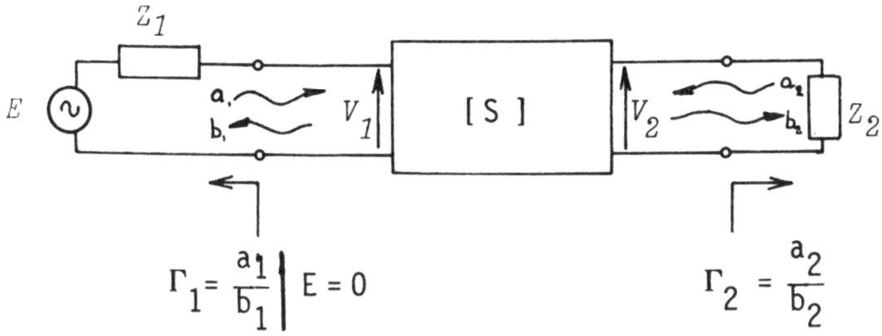

Fig. A1 *Reflection coefficients seen by the two-port network and leading to the flowchart of Fig. A2.*

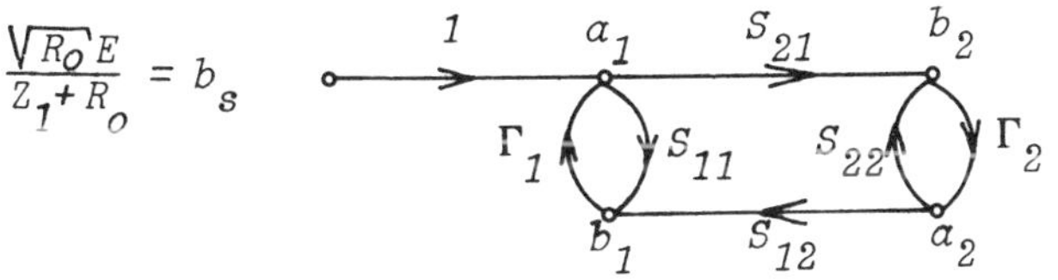

Fig. A2 *Flowchart of a source–two-port-network–load configuration.*

We now propose to calculate the voltage gain V_2/V_1. Using the relations

$$a_j = \frac{V_j + R_0 I_j}{2R_0^{\frac{1}{2}}} \quad \text{and} \quad b_j = \frac{V_j - R_0 I_j}{2R_0^{\frac{1}{2}}}$$

we obtain $V_j = (a_j + b_j)R_0^{\frac{1}{2}}$ and hence

$$\frac{V_2}{V_1} = \frac{a_2 + b_2}{a_1 + b_1}$$

which can be rewritten as

$$\frac{V_2}{V_1} = \frac{a_2/b_s + b_2/b_s}{a_1/b_s + b_1/b_s}$$

where $b_s = R_0^{\frac{1}{2}}E/(Z_1 + R_0)$.

We therefore have to calculate four 'transmittances'. We shall use Mason's rule, which is also known as the non-adjacent loop rule and is formulated as follows:

$$T = \frac{P_1(1 - \Sigma L_1{}^1 + \Sigma L_2{}^2 - \ldots) + P_2(1 - \Sigma L_1{}^2 + \Sigma L_2{}^2 - \ldots) + \ldots}{1 - \Sigma L_1 + \Sigma L_2 - \Sigma L_3 + \ldots \Sigma L_i} \quad \text{(A1)}$$

The denominator of this expression characterizes the flowchart alone and L_i represents an ith-order loop. Thus, for the chart in our example the first-order loops are $\Gamma_1 S_{11}$ and $\Gamma_2 S_{22}$, but also $\Gamma_1 S_{21} \Gamma_2 S_{12}$ (known as loop gain). Hence

$$\Sigma L_1 = \Gamma_1 S_{11} + \Gamma_2 S_{22} + \Gamma_1 S_{21} \Gamma_2 S_{12}$$

The second-order loops are obtained by taking the products of two non-adjacent first-order loops, i.e. loops without a common node a_1 or b_i; in the present case there is only one loop of this type and consequently

$$\Sigma L_2 = \Gamma_1 S_{11} \Gamma_2 S_{22}$$

The third-order loops would be obtained by taking the products of three non-adjacent loops, of which there are none at all in our example.

We now consider the numerator. We see that it is the weighted sum of the various paths connecting two selected variables. The weighting involves the terms L_i^j which represent the ith order loops non-adjacent to the jth path. Starting from b_s, there is only one path $P_1 = \Gamma_2 S_{21}$ to a_2 and all the loops are adjacent to it. There is likewise only one path $P_1 = S_{21}$ to b_2, and all the loops are adjacent to it. There is also only one path to a_1, but now there is a first-order non-adjacent loop $\Gamma_2 S_{22}$ and the 'transmittance' numerator is therefore $1(1 - \Gamma_2 S_{22})$.

The numerator N of b_1/b_s is more useful because there we have two possible paths, i.e. $P_1 = S_{11}$ and $P_2 = S_{21} \Gamma_2 S_{12}$; since for the first path there is one non-adjacent loop $\Gamma_2 S_{22}$, the numerator can be expressed in the form

$$N = S_{11}(1 - \Gamma_2 S_{22}) + S_{12} S_{21} \Gamma_2$$

The common denominator of each of the four transmittances which was given by

$$D = 1 - (\Gamma_1 S_{11} + \Gamma_2 S_{22} + \Gamma_1 S_{21} \Gamma_2 S_{12}) + \Gamma_1 S_{11} \Gamma_2 S_{22}$$

disappears in the ratio

$$\frac{a_2/b_s + b_2/b_s}{a_1/b_s + b_1/b_s}$$

The voltage gain is therefore

$$\frac{V_2}{V_1} = \frac{S_{21}(1+\Gamma_2)}{1(1-\Gamma_2 S_{22}) + S_{11}(1-\Gamma_2 S_{22}) + S_{12}S_{21}\Gamma_2}$$

which can be rewritten as

$$\frac{V_2}{V_1} = \frac{S_{21}(1+\Gamma_2)}{(1-\Gamma_2 S_{22})(1+S'_{11})} \tag{A2}$$

where

$$S'_{11} = S_{11} + \frac{S_{12}S_{21}\Gamma_2}{1-\Gamma_2 S_{22}}$$

S'_{11} is also the ratio b_1/a_1 when $\Gamma_1 = 0$.

Let us now only consider 'transmittance' b_1/b_s. It is equal to N/D, and we note that it can be written in the simplified form

$$\frac{b_1}{b_s} = \frac{S'_{11}}{1-\Gamma_1 S'_{11}} \tag{A3}$$

after dividing the top and bottom by $1 - \Gamma_2 S_{22}$.

Appendix B

Microstrip lines

B1. Synthesis formulae for microstrip lines

Formulae particularly suitable for alumina-type substrates, i.e. substrates with ε_r about 10, enable us to calculate w/h and λ_m in terms of Z_c and ε_r. The expressions given below are mainly due to the work of Owens, following Wheeler.

CALCULATION OF W/h

Case (a): $Z_c > (44 - 2\varepsilon_r)\,\Omega$

We introduce the intermediate quantity

$$h' = \frac{Z_c\{2(\varepsilon_r + 1)\}^{\frac{1}{2}}}{119.9} + \frac{1}{2}\frac{\varepsilon_r - 1}{\varepsilon_r + 1}\left\{\ln\left(\frac{\pi}{2}\right) + \frac{1}{\varepsilon_r}\ln\left(\frac{4}{\pi}\right)\right\}$$

and obtain

$$\frac{W}{h} = \left(\frac{\exp h'}{8} - \frac{1}{4\exp h'}\right)^{-1}$$

Case (b): $Z_c < (44 - 2\varepsilon_r)\,\Omega$

This time we introduce $D = 59.95\pi^2/Z_c\varepsilon_r$ which gives

$$\frac{W}{h} = \frac{2}{\pi}\{(D-1) - \ln(2D-1)\} + \frac{\varepsilon_r - 1}{\pi\varepsilon_r^{\frac{1}{2}}}\left\{0.293 - \frac{0.517}{\varepsilon_r} + \ln(D-1)\right\}$$

Using the charts of Fig. 3.21 we have plotted in Fig. B1 the strip widths W as a function of the relative permittivity ε_r for Z_c values of 20, 50, 100 and 200 Ω and an arbitrary height h of 1.59 mm (1/16 in). We have compared these with data obtained using Owens' formulae. The agreement is generally very good except when the characteristic impedance is about 50 Ω and ε_r is higher than 10. However, it does not appear critical to round up the value of 44 Ω at the boundary between cases (a) and (b), as shown by the redrawn curve at $Z_c = 50\,\Omega$. The W values can therefore be obtained using programs based on these formulae whatever the eventual value of ε_r.

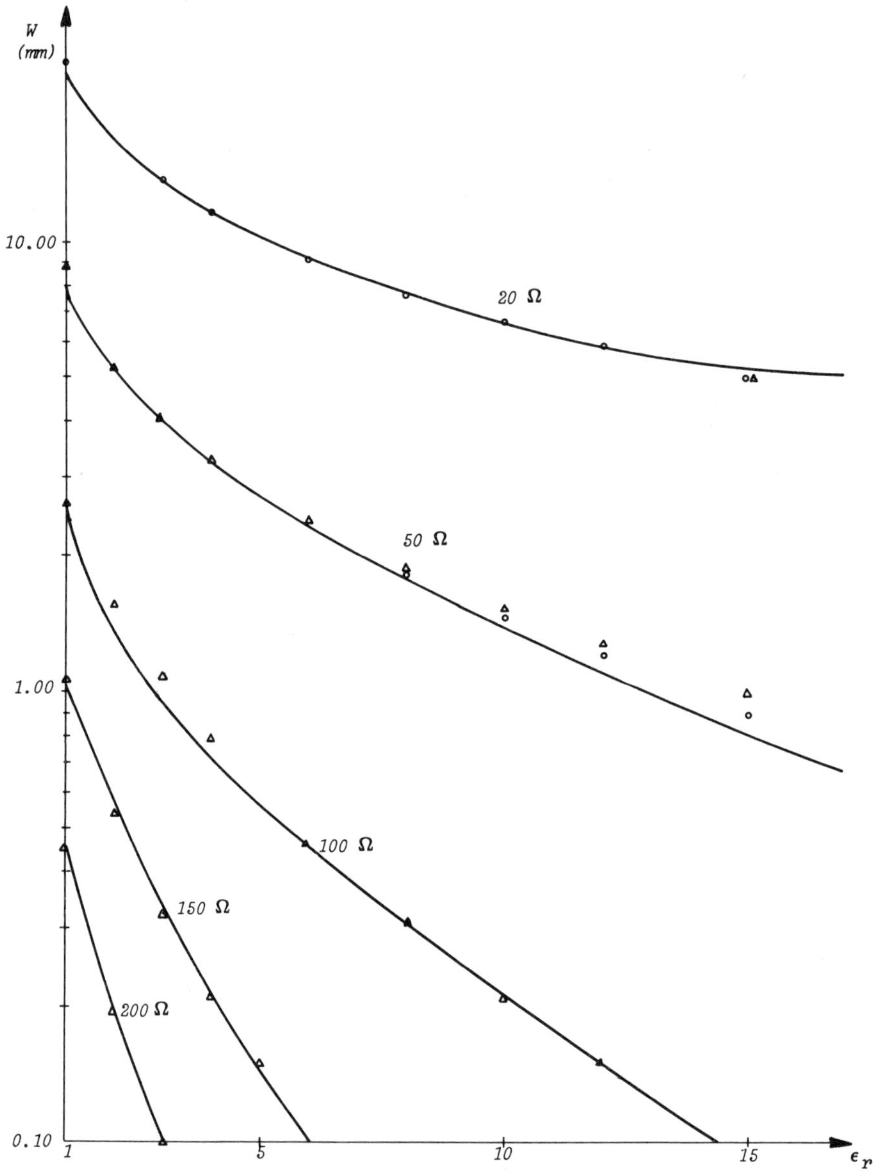

Fig. B1 *Variations in the strip width as a function of the relative permittivity ε_r for various characteristic impedances Z_c (20, 50, 100, 150 and 200 Ω) : ——, curves obtained from the charts of Fig. 3.21 ; △, case (a) ; ○, case (b).*

CALCULATION OF λ_m

The wavelength λ_m along the transmission structure can be derived from the free-space wavelength λ_0 by using the relation

$$\lambda_m = \frac{\lambda_0}{\varepsilon_{eff}^{\frac{1}{2}}}$$

where ε_{eff} is the effective permittivity of the microstrip line and is equal to ε_r for a dielectric-filled coaxial line. For a microstrip line in contact with a dielectric substrate on one side and air on the other, we have

$$\varepsilon_{eff} = 1 + q(\varepsilon_r - 1) \qquad \tfrac{1}{2} \leqslant q \leqslant 1$$

q is known as the **filling fraction** and its magnitude is proportional to the strip width. In practice we use the following formulae.

Case (a): $Z_c > (63 - 2\varepsilon_r)\,\Omega$
Again using the intermediate quantity h', we obtain

$$\varepsilon_{eff} = \frac{\varepsilon_r + 1}{2}\left[1 - \frac{1}{2h'}\left(\frac{\varepsilon_r - 1}{\varepsilon_r + 1}\right)\left\{\ln\left(\frac{\pi}{2}\right) + \frac{1}{\varepsilon_r}\ln\left(\frac{4}{\pi}\right)\right\}\right]^{-2}$$

Case (b): $Z_c < (63 - 2\varepsilon_r)\,\Omega$
We obtain directly

$$\varepsilon_{eff} = \frac{\varepsilon_r}{0.96 + \varepsilon_r(0.109 - 0.004\varepsilon_r)\{\lg(Z_c + 10) - 1\}}$$

Above a few gigahertz these expressions for ε_{eff} in the static TEM approximation have to be corrected as a function of the frequency f; hence

$$\lambda_m = \frac{\lambda_0}{\{\varepsilon_{eff}(f)\}^{\frac{1}{2}}}$$

where

$$\varepsilon_{eff}(f) = \varepsilon_r - \frac{\varepsilon_r - \varepsilon_{eff}}{1 + (h/Z_c)^{1.33}(0.43f^2 - 0.009f^3)}$$

(h is in millimetres and f is in gigahertz). The accuracy of this formula, which was proposed by Edwards and Owens, increases as the relative permittivity approaches that of alumina.

CORRECTION TERMS

Because of discontinuities corrections must be introduced, particularly at frequencies above 10 GHz.

For example, a $\lambda_m/8$ stub with one end in open circuit must be shortened by

$$\Delta l_0 = 0.412h\left(\frac{\varepsilon_{eff} + 0.3}{\varepsilon_{eff} - 0.258}\right)\left(\frac{W/h + 0.262}{W/h + 0.813}\right)$$

This empirical formula, which was given by Hammerstad and Bekkadal, can be applied to all ε_r values; Δl_0 represents the equivalent line length corresponding to the terminal parasitic capacitance ΔC_0, and we know from (3.9) that these two quantities are related by the identity

$$\frac{1}{\mathrm{j}\Delta C_0 \omega} \equiv \frac{Z_c}{\mathrm{j}\tan 2\pi \Delta l_0/\lambda_m}$$

If we connect two stubs of different widths W_1 and W_2 ($W_1 < W_2$) (Fig. B2), the wider stub must be shortened by approximately

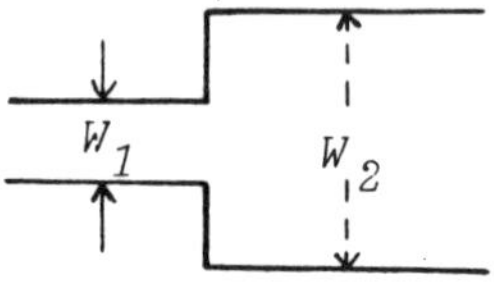

Fig. B2

$$\Delta l_2 \approx \Delta l_0 \left(1 - \frac{W_1}{W_2}\right)$$

for the same reason (terminal capacitance).

In the case of two stubs connected at right angles, we minimize the perturbations by intersecting the angle in such a way that $\theta \approx 55°$, as shown in Fig. B3.

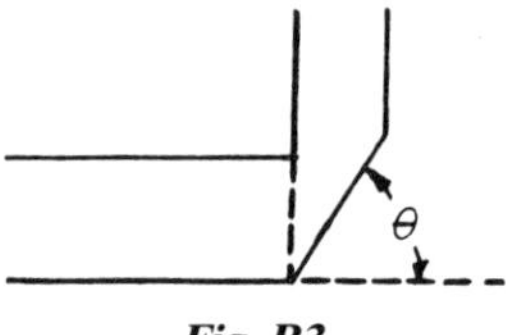

Fig. B3

B2. Dielectric substrates

Type of substrate	Relative permittivity ε_r	Notes on use
Dry air	1	The lines are suspended, which involves implementation constraints although these are not insurmountable; amplifiers have been designed up to several GHz
RT-Duroid 5880 (PTFE)	2.2	The dielectric (polytetrafluoroethylene) is reinforced with glass microfibres; it can be used up to very high frequencies (experiments at 100 GHz); other manufacturers offer this type of Teflon–fibreglass substrate with dielectric constants of 2.22, 2.33, 2.45, 2.55

Polyguide	2.32	Plastic usable at UHF but difficult to process; soft material whose deformation may cause lines to disintegrate (thickness, 1/16 in)
Quartz (SiO_2)	3.82	Its low ε_r makes it preferable to alumina above 20–30 GHz as the strips constructed are wider for a given impedance; virtually zero temperature coefficient; good conductor adhesion for the layer sequence Ta_2O_3/Ti/Pd/Au
Epoxy (glass fibre, resin)	4.9	Has replaced Bakelite for printed circuit fabrications; so widely used that the amateur can consider it for UHF applications (1/16 in double-sided epoxy)
Beryllium (BeO)	6.6	Note that this is a toxic product
Alumina (99.5% Al_2O_3)	9.8–10	Very widely employed between 2 and 20 GHz, but fragile and therefore only usually available in 2 in × 2 in wafers; good conductor bonding, giving the configurations Cr/Cu/(Pd or Pt)/Au; or Ti/(Pd or Pt)/Au
Sapphire (crystallized alumina)	10	Transparent and anisotropic from dielectric point of view; expensive
3M-Epsilam 10 (Teflon)	10.3	Recent product, competing with alumina, available in thicknesses of 1/20 in or 1/40 in
RT-Duroid 6010	10.5	PTFE and ceramic based; same frequency range as RT-Duroid 6006
GaAs	13	Used in microstrip and monolithic technologies
Ferrites	15	Magnetically anisotropic; useful where non-reciprocal effects are required
RT-Duroid 6006	6	PTFE and ceramic compound; can be used up to at least 10 GHz

The following apparent anomaly should be noted: the higher the frequency is, the thinner is the dielectric substrate (1/40 in (0.635 mm) and sometimes less at 10 GHz for alumina; 0.30 mm up to 50 GHz (0.15 mm extra for quartz); 1/100 in (0.254 mm) and then 1/200 in (127 μm) for applications up to 100 GHz on RT-Duroid 5880 substrates).

In order to avoid excessively small strip widths, which also means finding low relative permittivities, it is desirable to maintain a certain thickness.

Unfortunately this becomes impossible at high frequencies. In fact, the operating frequencies are then likely to be higher than the cut-off frequencies of the bulk modes inevitably present in microstrip and other structures. However, apart from the technical difficulties, reducing the substrate thickness poses the problem of power-handling capability. A final important limitation on the use of these structures at high frequencies is the existence of radiation losses and the suppression of unwanted modes.

Appendix C

Critical instability circles

C1. Generalization of the critical instability circle concept

We now attempt to find the locus of the reflection coefficients Γ_L such that

$$\left| S_{11} + \frac{S_{12}S_{21}\Gamma_L}{1 - S_{22}\Gamma_L} \right| = k$$

Reducing to a common denominator and introducing $\Delta = S_{11}S_{22} - S_{12}S_{21}$ gives

$$|S_{11} - \Delta\Gamma_L| = k|1 - \Gamma_L S_{22}|$$

which, on squaring, becomes

$$|S_{11}|^2 + |\Delta|^2|\Gamma_L|^2 - 2\,\mathrm{Re}(\Delta\Gamma_L S_{11}^*) = k^2 + k^2|\Gamma_L|^2|S_{22}|^2 - 2k^2\,\mathrm{Re}(\Gamma_L S_{22})$$

This can be rearranged to

$$(k^2|S_{22}|^2 - |\Delta|^2)|\Gamma_L|^2 - 2\,\mathrm{Re}\,\{(k^2 S_{22} - \Delta S_{11}^*)\Gamma_L\} = |S_{11}|^2 - k^2$$

Dividing by the quantity $k^2|S_{22}|^2 - |\Delta|^2$, which is assumed to be non-zero, we obtain

$$|\Gamma_L|^2 - \frac{2}{k^2|S_{22}|^2 - |\Delta|^2}\,\mathrm{Re}\,\{(k^2 S_{22} - \Delta S_{11}^*)\Gamma_L\} = \frac{|S_{11}|^2 - k^2}{k^2|S_{22}|^2 - |\Delta|^2}$$

This can easily be equated with the equation of a circle of centre Ω and radius R in the complex plane which is expressed in accordance with Eqn (2.19) as

$$|\Gamma|^2 - 2\,\mathrm{Re}(\Omega^*\Gamma) = R^2 - |\Omega|^2$$

We therefore obtain

$$\Omega^*(k) = \frac{k^2 S_{22} - \Delta S_{11}^*}{k^2|S_{22}|^2 - |\Delta|^2}$$

$$R^2(k) = \frac{|S_{11}|^2 - k^2}{k^2|S_{22}|^2 - |\Delta|^2} + \frac{(k^2 S_{22} - \Delta S_{11}^*)(k^2 S_{22} - \Delta S_{11}^*)^*}{(k^2|S_{22}|^2 - |\Delta|^2)^2}$$

We then reduce the expression for $R^2(k)$ to the same denominator. After expansion the numerator becomes

$$N^2 = k^2|S_{11}|^2|S_{22}|^2 - |\Delta|^2|S_{11}|^2 - k^4|S_{22}|^2 + k^2|\Delta|^2 + \\ + k^4|S_{22}|^2 + |\Delta|^2|S_{11}|^2 - k^2(\Delta^*S_{11}S_{22} + \Delta S_{11}{}^*S_{22}{}^*)$$

which simplifies to

$$N^2 = k^2|S_{11}|^2|S_{22}|^2 + k^2|\Delta|^2 - k^2(\Delta^*S_{11}S_{22} + \Delta S_{11}{}^*S_{22}{}^*)$$

Expanding the term in parentheses gives

$$\Delta^*S_{11}S_{22} + \Delta S_{11}{}^*S_{22}{}^* = 2|S_{11}|^2|S_{22}|^2 - S_{12}{}^*S_{21}{}^*S_{11}S_{22} - S_{12}S_{21}S_{11}{}^*S_{22}{}^*$$

Now

$$|\Delta|^2 = |S_{11}|^2|S_{22}|^2 - S_{12}S_{21}S_{11}{}^*S_{22}{}^* - S_{12}{}^*S_{21}{}^*S_{11}S_{22} + |S_{12}|^2|S_{21}|^2$$

and therefore

$$\Delta^*S_{11}S_{22} + \Delta S_{11}{}^*S_{22}{}^* = |S_{11}|^2|S_{22}|^2 - |S_{12}|^2|S_{21}|^2 + |\Delta|^2$$

Substituting in the expression for the numerator, we simply have

$$N^2 = k^2|S_{12}|^2|S_{21}|^2$$

Finally, the required locus is the circle C_k with centre

$$\Omega(k) = \frac{(k^2S_{22} - \Delta S_{11}{}^*)^*}{k^2|S_{22}|^2 - |\Delta|^2}$$

and radius

$$R(k) = \frac{k|S_{12}|\,|S_{21}|}{|k^2|S_{22}|^2 - |\Delta|^2|}$$

C2. Locus of the centres of the instability circles

We calculate the angle $\widehat{OPK}$ where O is the origin of the real and imaginary axes, P is the point representing $\Omega(1)$ and K is the point representing $\Omega(k)$ ($k > 1$). For this purpose, we use the triangular relation

$$|\Omega(k)|^2 = |\Omega(1)|^2 + |\Omega(k) - \Omega(1)|^2 - 2|\Omega(1)|\,|\Omega(k) - \Omega(1)| \cos\theta_1$$

in which $\theta_1 = \widehat{OPK}$. Next, we calculate

$$|\Omega(k)|^2 = \frac{k^4|S_{22}|^2 + |\Delta|^2|S_{11}|^2 - k^2(\Delta^*S_{11}S_{22} + \Delta S_{11}{}^*S_{22}{}^*)}{(k^2|S_{22}|^2 - |\Delta|^2)^2}$$

The term in parentheses in the numerator has already been calculated in Section C1, and after rearranging we find

$$|\Omega(k)|^2 = \frac{(k^2|S_{22}|^2 - |\Delta|^2)(k^2 - |S_{11}|^2) + k^2|S_{12}|^2|S_{21}|^2}{(k^2|S_{22}|^2 - |\Delta|^2)^2}$$

In particular,

$$|\Omega(1)|^2 = \frac{(|S_{22}|^2 - |\Delta|^2)(1 - |S_{11}|^2) + |S_{12}|^2|S_{21}|^2}{(|S_{22}|^2 - |\Delta|^2)^2}$$

After reducing

$$\Omega(k) = \frac{k^2 S_{22}{}^* - \Delta^* S_{11}}{k^2|S_{22}|^2 - |\Delta|^2}$$

and

$$\Omega(1) = \frac{S_{22}{}^* - \Delta^* S_{11}}{|S_{22}|^2 - |\Delta|^2}$$

to the same denominator we obtain

$$\Omega(k) - \Omega(1) = \frac{(k^2 - 1)S_{12}S_{21}\Delta^* S_{22}{}^*}{(k^2|S_{22}|^2 - |\Delta|^2)(|S_{22}|^2 - |\Delta|^2)}$$

Hence

$$|\Omega(k) - \Omega(1)|^2 = \frac{(k^2 - 1)^2|S_{12}|^2|S_{21}|^2|\Delta|^2|S_{22}|^2}{(k^2|S_{22}|^2 - |\Delta|^2)^2(|S_{22}|^2 - |\Delta|^2)^2}$$

The triangular relation is formulated as

$$2|\Omega(1)|\cos\theta_1 = \frac{|\Omega(1)|^2 - |\Omega(k)|^2 + |\Omega(k) - \Omega(1)|^2}{|\Omega(k) - \Omega(1)|}$$

Consider the quantity

$$|\Omega(1)|^2 - |\Omega(k)|^2 + |\Omega(k) - \Omega(1)|^2$$

Its common denominator is

$$D = (k^2|S_{22}|^2 - |\Delta|^2)^2(|S_{22}|^2 - |\Delta|^2)^2$$

The numerator is therefore

$$\begin{aligned} N = {} & \{(|S_{22}|^2 - |\Delta|^2)(1 - |S_{11}|^2) + |S_{12}|^2|S_{21}|^2\}(k^2|S_{22}|^2 - |\Delta|^2)^2 - \\ & - \{(k^2|S_{22}|^2 - |\Delta|^2)(k^2 - |S_{11}|^2) + k^2|S_{12}|^2|S_{21}|^2\}(|S_{22}|^2 - |\Delta|^2)^2 + \\ & + (k^2 - 1)^2|S_{12}|^2|S_{21}|^2|\Delta|^2|S_{22}|^2 \end{aligned}$$

We put it in the form $N = N_1 + N_2$ with

$$\begin{aligned} N_1 = {} & (k^2|S_{22}|^2 - |\Delta|^2)^2(|S_{22}|^2 - |\Delta|^2)(1 - |S_{11}|^2) - \\ & - (k^2|S_{22}|^2 - |\Delta|^2)(|S_{22}|^2 - |\Delta|^2)^2(k^2 - |S_{11}|^2) \end{aligned}$$

$$= (k^2|S_{22}|^2 - |\Delta|^2)(|S_{22}|^2 - |\Delta|^2)\{(k^2|S_{22}|^2 - |\Delta|^2)(1 - |S_{11}|^2) - - (|S_{22}|^2 - |\Delta|^2)(k^2 - |S_{11}|^2)\}$$

$$= (1 - k^2)(k^2|S_{22}|^2 - |\Delta|^2)(|S_{22}|^2 - |\Delta|^2)(|S_{11}|^2 - |S_{22}|^2 - |\Delta|^2)$$

$$N_2 = |S_{12}|^2|S_{21}|^2\{(k^2|S_{22}|^2 - |\Delta|^2)^2 - k^2(|S_{22}|^2 - |\Delta|^2)^2 + (k^2 - 1)^2|\Delta|^2|S_{22}|^2\}$$

$$= (k^2 - 1)|S_{12}|^2|S_{21}|^2\{k^2|S_{22}|^4 - |\Delta|^4 + (k^2 - 1)|S_{22}|^2|\Delta|^2\}$$

$$= (k^2 - 1)|S_{12}|^2|S_{21}|^2(k^2|S_{22}|^2 - |\Delta|^2)(|S_{22}|^2 + |\Delta|^2)$$

We thus obtain

$$\frac{N_1 + N_2}{D} = \frac{(k^2 - 1)\{(|S_{22}|^2 - |\Delta|^2)(|\Delta|^2 - |S_{11}|^2|S_{22}|^2) + (|S_{22}|^2 + |\Delta|^2)|S_{12}|^2|S_{21}|^2\}}{(k^2|S_{22}|^2 - |\Delta|^2)(|S_{22}|^2 - |\Delta|^2)^2}$$

and on dividing by the modulus $|\Omega(k) - \Omega(1)|$, the square of which was calculated above, we have the relation from which we obtain θ_1:

$$2|\Omega(1)|\cos\theta_1 = \frac{(|S_{22}|^2 - |\Delta|^2)(|\Delta|^2 - |S_{11}|^2|S_{22}|^2) + (|S_{22}|^2 + |\Delta|^2)|S_{12}|^2|S_{21}|^2}{(|S_{22}|^2 - |\Delta|^2)|S_{12}||S_{21}||\Delta||S_{22}|}$$

This expression for θ_1 is independent of k; the centres of the instability circles are therefore aligned and the representative straight line can be drawn directly when we have located the point $\Omega(1)$.

C3. Tangential instability circle

The problem consists of determining the value of k for which the circle C_k is tangential to the circumference of the Smith chart (in the case where $|K|$ is less than unity).

When $|\Delta|$ is less than $|S_{22}|$ and $|S_{22}|$ is less than unity, there exists a particular value k_t such that the circle C_{k_t} is externally tangential to the Smith chart. We then have the equality

$$|\Omega(k_t)| = R(k_t) + 1$$

which, on squaring, gives

$$(k_t^2|S_{22}|^2 - |\Delta|^2)(k_t^2 - |S_{11}|^2) + k_t^2|S_{12}|^2|S_{21}|^2 = \{k_t|S_{12}||S_{21}| + (k_t^2|S_{22}|^2 - |\Delta|^2)\}^2$$

Expansion of the terms in parentheses on the right-hand side in the form of a sum of two terms gives

$$(k_t^2|S_{22}|^2 - |\Delta|^2)(k_t^2 - |S_{11}|^2) = (k_t^2|S_{22}|^2 - |\Delta|^2)(k_t^2|S_{22}|^2 - |\Delta|^2 + 2k_t|S_{12}||S_{21}|)$$

This leads to the following second-degree expression in k_t:

$$(1-|S_{22}|^2)k_t^2 - 2k_t|S_{12}|\,|S_{21}| + |\Delta|^2 - |S_{11}|^2 = 0$$

The solution adopted is

$$k_t = \frac{|S_{12}|\,|S_{21}| + \{|S_{12}|^2|S_{21}|^2 - (1-|S_{22}|^2)(|\Delta|^2 - |S_{11}|^2)\}^{\frac{1}{2}}}{1-|S_{22}|^2}$$

In fact, we can show that for $|K| = 1$ we have, from Eqn (2.14),

$$|S_{12}|^2|S_{21}|^2 = \frac{\{(1-|S_{22}|^2) + (|\Delta|^2 - |S_{11}|^2)\}^2}{4}$$

The root is therefore equal to

$$\frac{(1-|S_{22}|^2) - (|\Delta|^2 - |S_{11}|^2)}{2}$$

and hence $k_t = 1$ as required.

When $|\Delta|$ is greater than $|S_{22}|$, the formula giving k_t is still valid; the equality to be satisfied is $|\Omega(k_t)| = R(k_t) - 1$, but then the denominator of $R(k_t)$ to be considered is $|\Delta|^2 - k_t^2|S_{22}|^2$.

Note

The expression giving k_t can be written in a more concise form because

$$|S_{12}|^2|S_{21}|^2 - (1-|S_{22}|^2)(|\Delta|^2 - |S_{11}|^2) = |C_1|^2$$

(see the derivation of Eqn (2.14)). Hence

$$k_t = \frac{|S_{12}|\,|S_{21}| + |C_1|}{1-|S_{22}|^2}$$

Appendix D

Allan's variance and spectral density

Transition from the time domain variance measurement to the frequency domain spectral density measurement

Using the expression

$$\sigma_{\bar{y}}^2(N, T, \tau) = \frac{1}{N}\sum_{n=1}^{N} \bar{y}_n^{\,2} - \left(\frac{1}{N}\sum_{n=1}^{N} \bar{y}_n\right)^2$$

in which

$$\bar{y}_n = \frac{1}{2\pi f_0}\frac{\Delta\phi_n}{\tau}$$

we obtain

$$\sigma_{\bar{y}}^2(N, T, \tau) = \frac{1}{(2\pi f_0\tau)^2}\sigma_{\Delta\phi}^{\;2}(N, T, \tau)$$

Experimentally, M series of N measurements are performed, and for the kth series, for example, we determine the quantity

$$\sigma_k^{\,2}(N, T, \tau) = \frac{1}{N}\sum_{n=1}^{N}\frac{\Delta\phi_n^{\,2}}{\tau^2} - \left(\frac{1}{N}\sum_{n=1}^{N}\frac{\Delta\phi_n}{\tau}\right)^2$$

The instrument will then provide a statistical mean:

$$E\{\sigma_k^{\,2}(N, T, \tau)\} = \frac{1}{M}\sum_{k=1}^{M}\sigma_k^{\,2}(N, T, \tau)$$

Provided that the ergodicity of the process is established, we have the identity

$$E\{\sigma_k^{\,2}(N, T, \tau)\} \equiv \overline{\sigma_k^{\,2}(N, T, \tau)}$$

Expanding the right-hand side gives

$$\overline{\sigma_k^{\,2}(N, T, \tau)} = \frac{1}{N}\sum_{n=1}^{N}\frac{\overline{\Delta\phi_n^{\,2}}}{\tau^2} - \frac{1}{\tau^2 N^2}\overline{\left(\sum_{n=1}^{N}\Delta\phi_n\right)\left(\sum_{p=1}^{N}\Delta\phi_p\right)}$$

which can be rewritten as

$$\overline{\sigma_k{}^2(N, T, \tau)} = \frac{1}{\tau^2}\left(\frac{1}{N}\sum_{n=1}^{N}\overline{\Delta\phi_n{}^2} - \frac{1}{N^2}\sum_{n=1}^{N}\sum_{p=1}^{N}\overline{\Delta\phi_n\Delta\phi_p}\right)$$

where

$$\overline{\Delta\phi_n\Delta\phi_p} = \overline{\{\phi(nT+\tau) - \phi(nT)\}\{\phi(pT+\tau) - \phi(pT)\}}$$

Taking the product, we obtain the autocorrelation function Γ_ϕ of $\phi(t)$, and hence

$$\overline{\Delta\phi_n\Delta\phi_p} = 2\Gamma_\phi\{(n-p)T\} - \Gamma_\phi\{(n-p)T - \tau\} - \Gamma_\phi\{(n-p)T + \tau\}$$

or in a more concise form

$$\overline{\Delta\phi_n\Delta\phi_p} = \Gamma_{\phi_{np}}(\tau)$$

We note that for $n = p$ we obtain

$$\overline{\Delta\phi_n{}^2} = 2\{\Gamma_\phi(0) - \Gamma_\phi(\tau)\}$$

The time domain mean of the variance can therefore be expressed in terms of Γ_ϕ:

$$\overline{\sigma_k{}^2(N, T, \tau)} = \frac{1}{\tau^2}\left\{2\Gamma_\phi(0) - 2\Gamma_\phi(\tau) - \frac{1}{N^2}\sum_{n=1}^{N}\sum_{p=1}^{N}\Gamma_{\phi_{np}}(\tau)\right\}$$

Now, the part of the double sum for which $n = p$ is given by $2N\{\Gamma_\phi(0) - \Gamma_\phi(\tau)\}$. If we take the symmetry of the autocorrelation function into account, we obtain

$$\overline{\sigma_k{}^2(N, T, \tau)} = \frac{2}{\tau^2}\left[\left(1 - \frac{1}{N}\right)\{\Gamma_\phi(0) - \Gamma_\phi(\tau)\} - \frac{1}{N^2}\sum_{n=2}^{N}\sum_{p=1}^{n-1}\Gamma_{\phi_{np}}(\tau)\right]$$

If we now want to obtain the power spectral density $G_\phi(f)$, it is necessary to take the Fourier transform of $\tau^2\overline{\sigma_k{}^2(N, T, \tau)}$. It should be noted that

$$\text{FT}\{\Gamma_\phi(\tau)\} = G_\phi(f)$$

and

$$\text{FT}[\Gamma_\phi\{(n-p)T \pm \tau\}] = \exp\{\pm i2\pi f(n-p)T\}G_\phi(f)$$

which enables us to write, excluding the zero-frequency deviation,

$$\text{FT}\{\tau^2\overline{\sigma_k{}^2(N, T, \tau)}\} = \left[-2\left(1 - \frac{1}{N}\right) + \frac{4}{N^2}\sum_{n=2}^{N}\sum_{p=1}^{n-1}\cos\{2\pi f(n-p)T\}\right]G_\phi(f)$$

The double sum is calculated from the complex sum of the (general) term $\exp\{i(n-p)\theta\}$ ($\theta = 2\pi fT$) and we finally obtain

$$\text{FT}\{\tau^2\overline{\sigma_k{}^2(N, T, \tau)}\} = -2\left\{1 - \frac{1}{N^2}\frac{1 - \cos(2\pi fNT)}{1 - \cos(2\pi fT)}\right\}G_\phi(f)$$

It is therefore possible to use the time domain variance measurement to obtain the spectral density $G_{\phi}(f)$. If we take the equalities

$$\tau^2 \sigma_k{}^2(N, T, \tau) = \sigma_{\Delta\phi}{}^2(N, T, \tau) = (2\pi f_0 \tau)^2 \sigma_{\bar{y}}{}^2(N, T, \tau)$$

into account the transition relation will be

$$\mathrm{FT}\left\{(2\pi f_0 \tau)^2 \frac{1}{M} \sum_{k=1}^{M} \sigma_{\bar{y}}{}^2(N, T, \tau)\right\} = -2\left\{1 - \frac{1}{N^2} \frac{1 - \cos(2\pi f N T)}{1 - \cos(2\pi f T)}\right\} G_{\phi}(f)$$

Bibliography

ALLAN D. W. (1966): Statistics of atomic frequency standards, *Proceedings of the IEEE*, **54** (2), 221–230.

ANDERSON R. W. (February 1967): *S*-parameter techniques for faster, more accurate network design, *Hewlett Packard Journal*; HEWLETT PACKARD, *Application Note 95–1*.

ANGOT A. (1982): *Compléments de Mathématiques*, 2nd edn, Masson, Paris.

BADOUAL R. (1983): *Les Micro-ondes*, Masson, Paris (two volumes).

BARNES J. A. *et al.* (1971): Characterization of frequency stability, *IEEE Transactions on Instrumentation and Measurements*, **20** (2), 105–120.

BERTEAUD A. J. (1976): *Les Hyperfréquences*, Presses Universitaires de France (*Que sais-je no. 1643*), Paris.

BLANC-LAPIÈRRE A. and PICINBONO B. (1981): *Fonctions Aléatoires*, Masson, Paris.

BOUDOURIS G. and CHENEVIER P. (1975): *Circuits pour Ondes Guidées*, Dunod, Paris.

CARSON R. S. (1982): *High-frequency Amplifiers*, 2nd edn, Wiley, New York.

CAULTON M. (1979): Substrates, materials and processes for microwave applications. *Proceedings of the 29th Electronic Components Conf.*, pp. 126–131, IEEE, Piscataway, NJ.

COMBES P. F. (1980): *Ondes Métriques et Centimétriques: Lignes, Circuits Passifs, Antennes*, Dunod, Paris.

COMBES P. F. (1983): *Transmission en Espace Libre et sur les Lignes*, Dunod, Paris.

DUBOST G. (1981): *Propagation Libre et Guidée des Ondes Electro-magnétiques, Rayonnement*, Masson, Paris.

EDWARDS T. C. (1983): *Conception des Circuits Micro-ondes* (*Foundations for Microstrip Circuit Design*) (translated by GREZAUD J. and TEISSON J.), Masson, Paris.

GARDIOL F. (1981): Hyperfréquences, *Traité d'Electricité*, **13**, Georgi, Lausanne.

GRIVET P. (1969): *Physique des Lignes de Haute Fréquence et d'Ultra-haute Fréquence*, Vol. 1, *Paramètres Primaires et Secondaires, Ondes Progressives, Impulsions*, Masson, Paris.

GRIVET P. (1974): *Physique des Lignes de Haute Fréquence et d'Ultra-haute Fréquence*, Vol. 2, *Circuits et Amplificateurs Micro-ondes*, Masson, Paris (two fascicules).

GUPTA K. C., GARG R., RAMESH G. and BAHL I. J. (1979): *Microstrip Lines and Slotlines*, Artech House, Dedham, MA.

HAMMERSTAD E. O. (1975): Equations for microstrip circuit design. *Proceedings of the 5th European Microwave Conf., Hamburg*, pp. 268–272.

HAMMERSTAD E. O. and BEKKADAL F. (1975): *A Microstrip Handbook, ELAB Report STF 44 A 74169, N 7034*, Norwegian Institute of Technology, University of Trondheim.

HERVÉ J. (1981): *Electronique Appliquée à la Transmission de l'Information*, Masson, Paris (two volumes).

HEWLETT PACKARD (April 1972): *S*-parameter design, *Application Note 154*.

Hewlett Packard (June 1980): Automating the HP 8410 B microwave network analyzer, *Application Note 221 A*.

Hewlett Packard (July 1983): Principes fondamentaux des mesures de facteur de bruit en radiofréquences et en hyperfréquences, *Note d'Applications 57–1*.

Higashisaka A. and Mizuta T. (1981): 20-GHz band monolithic GaAs FET low-noise amplifier, *IEEE Transactions on Microwave Theory and Techniques*, **29** (1), 1–6.

Howe H. (1979): *Stripline Circuit Design*, 3rd edn, Artech House, Dedham, MA.

Kline M. D. and Nelson T. M. (1980): Low cost large substrates for microwave integrated circuits. *Proc. Int. Microelectronics Symp., New York*, pp. 81–88, Int. Soc. for Hybrid Microelectronics, Montgomery, AL.

Kurokawa K. (1965): Power waves and the scattering matrix, *IEEE Transactions on Microwave Theory and Techniques*, **13** (2), 194–202.

Lefeuvre S. (1969): *Hyperfréquence*, Dunod, Paris.

Matthaei G. L., Young L. and Jones E. M. T. (1980): *Microwave Filters, Impedance-matching Networks, and Coupling Structures*, Artech House, Dedham, MA.

Owen R. P. (1976): Accurate analytical determination of quasistatic microstrip line parameters, *The Radio and Electronic Engineer*, **46** (7), 360–364.

Picinbono B. (1980): *Eléments de Théorie du Signal*, Dunod, Paris.

Rivier E. and Sardos R. (1982): *La Matrice S: du Numérique à l'Optique*, Masson, Paris.

Saad T. S. (1971): *Microwave Engineer's Handbook*, Artech House, Dedham, MA (two volumes).

Strid E. (1981): Noise measurements for low-noise GaAs FET amplifiers, *Microwave Systems News*, **11** (11), 62–70.

Strid E. (1981): Noise measurement check list eliminates costly errors, *Microwave Systems News*, **11** (12), 88–107.

Vendelin G. D. (1982): *Design of Amplifiers and Oscillators by the S-parameter Method*, Wiley, New York.

Wheeler H. A. (1965): Transmission line properties of parallel strips separated by a dielectric sheet, *IEEE Transactions on Microwave Theory and Techniques*, **13** (3), 172–185.

Wheeler H. A. (1977): Transmission line properties of a strip on a dielectric sheet on a plane, *IEEE Transactions on Microwave Theory and Techniques*, **25**, 631–647.

Index